D0152148

DEVELOPING ALBERTA'S OIL SANDS

from Karl Clark to Kyoto

DEVELOPING ALBERTA'S OIL SANDS

from Karl Clark to Kyoto

paul chastko

UNIVERSITY OF
CALGARY
PRESS

Published by the University of Calgary Press
2500 University Drive NW, Calgary, Alberta, Canada T2N 1N4
www.uofcpress.com

This book has been published with the help of a grant from the Canadian Federation
for the Humanities and Social Sciences, through the Aid to Scholarly Publications
Programme, using funds provided by the Social Sciences and Humanities Research
Council of Canada.

The University of Calgary Press acknowledges the support of the Alberta Foundation of
the Arts for this published work.

We acknowledge the financial support of the Alberta Lottery Fund -- Community
Initiatives program.

Canada Council   Conseil des Arts
for the Arts   du Canada

ALBERTA
LOTTERY FUND

We acknowledge the financial support of the Government of Canada through the Book
Publishing Industry Development Program (BPIDIP) for our publishing activities.

National Library of Canada Cataloguing in Publication

Chastko, Paul A. (Paul Anthony)
     Developing Alberta's oil sands: from Karl Clark to Kyoto /
Paul A. Chastko.

Includes bibliographical references and index.
ISBN 1-55238-124-2

     1. Athabasca Tar Sands (Alta.) – History. 2. Oil sands
industry – Alberta – History. 3. Oil sands – Alberta – History.
I. Title.

HD9574.C23A54 2004   338.2'7283'097123   C2004-901361-0

∞ This book is printed on acid-free paper

Printed and bound in Canada by Friesens

Cover design by Mieka West
Page design and typesetting by Elizabeth Gusnoski

CONTENTS

LIST OF TABLES

LIST OF MAPS

LIST OF ABBREVIATIONS

AEC	Alberta Energy Company
AOSTRA	Alberta Oil Sands Technical Research Authority
AOSERP	Alberta Oil Sands Environmental Research Program
APG	Alsands Project Group
API	American Petroleum Institute
CAPP	Canadian Association of Petroleum Producers
CDC	Canadian Development Corporation
CM&S	Canadian Mining and Smelting Limited
CNRL	Canadian Natural Resources Limited
CONRAD	Canadian Oil Sands Network for Research & Development

CPA	Canadian Petroleum Association
DET	Domestic Emissions Trading
DOE	U.S. Department of Energy
ERCB	Alberta Energy Resources and Conservation Board
EMR	Federal Department of Energy, Mines and Resources
EnFin	Federal Energy and Finance Portfolio
EUB	Alberta Energy Utilities Board
EU	European Union
FIRA	Foreign Investment Review Agency
FTA	Canada-United States Free Trade Agreement
GA	Glenbow Archives
GCOS	Great Canadian Oil Sands Limited
GHG	Greenhouse Gasses
IEA	International Energy Agency
IORT	Incremental Oil Revenue Tax
IPAC	Independent Petroleum Association of Canada
LNG	Liquefied Natural Gas
MNC	Multinational Corporation
NAFTA	North American Free Trade Agreement
NEB	National Energy Board
NEP	National Energy Program
NORP	New Oil Reference Price
O/A	Office of the Assistant
OGCB	Alberta Oil and Gas Conservation Board
OPEC	Oil Producing Exporting Countries
OSLO	Other Six Lease Operators
PAA	Provincial Archives of Alberta
PGRT	Petroleum and Gas Revenue Tax
PIP	Petroleum Incentives Program
SAGD	Steam Assisted Gravity Drainage
SMB	Sidney Martin Blair Papers
TSE	Toronto Stock Exchange
UAA	University of Alberta Archives
WHCF	White House Central Files

ACKNOWLEDGMENTS

The most rewarding part about producing a book comes from finally being able to repay the multitude of debts incurred during years of study, research, and writing. First and foremost, I would like to thank Walter Hildebrandt, the Director of University of Calgary Press and the editorial staff at UC Press – Joan Barton, John King, Mieka West, and Dianne Hiebert – for taking on the project and guiding it toward publication. I would also like to thank my dissertation supervisor, Dr. Alfred E. Eckes, Jr. for all his support and encouragement, both during and after my studies at Ohio University. Every graduate student should be so lucky as I am to have such a wonderful and encouraging mentor. This book would not have been produced if it were not for his keen eye and willingness to hear me tell him about "something we have back in Alberta called the oil sands."

I would also like to thank the people who took the time to read and comment on the manuscript in its various forms. The members of my dissertation committee, Dr. Alonzo Hamby, Dr. Richard Vedder, Dr. Peter John Brobst who read an earlier draft, and the anonymous reviewers at University of Calgary Press. Their questions, comments, opinions, and suggestions have undoubtedly made the finished product much stronger. In addition to the members of my dissertation committee mentioned above, professors Bruce Steiner, Steven Miner, Norman Goda, Joan Hoff, Jeffrey Herf, Chester Pach, and Marvin Fletcher encouraged and prodded me to become a better writer and historian. I hope they see in this final product the results of their efforts. Of course, any errors of judgment or interpretation that may remain, however, are my own.

The staff and students of Ohio University's Department of History and the Contemporary History Institute made my time as a graduate student in Ohio pass all too quickly. Both the Department of History and the Contemporary History Institute provided grants to complete the research on this book. Special thanks also go out to the archival staff of the depositories listed in the bibliography. In particular, I would like to thank Albert Nason at the Jimmy Carter Presidential Library, Angela Jones at the Alberta Energy and Utilities Board, and Doug Cass of the Glenbow Archives, all of whom greeted my inquiries with smiles and good humour. Without the assistance of these people, the task of researching this dissertation would have been much more difficult.

I could probably fill an additional three hundred pages about the encouragement and much-needed mirth provided by friends across North America and beyond. Morten Bach, Scott Beekman, Ruth "Grammie" Boutilier, Rob and Shelby Burns, Dave Chin, Korcaighe Hale, George and Judy Glowa, Tim Glowa and Wendy Whamond, Chris Jackson, Bill Kamil, Chris and Jackie Kane, Kim Little, Scott McCleary, Kirk Tyvela, Stephan Rothlander, Jim and Joanne Wachal, Richard and Wanda Wachal, Randy and Nadine Wachal, Gord and Michelle Wilks, and Matt and Penni Zimmerman all offered words of encouragement and helped distract me when distractions were needed.

I used to think that there might be a special place in the hereafter reserved for the family of doctoral students and academic writers. My wife, Michelle, assures me there is no need to wonder any longer – all anyone would have to do is look at the stack of books, papers, and notes on my desk (and on the floor ... or in the living room ... or even the small pile in the kitchen) to conclude that she has already served her time in purgatory. She is, of course, right. Families tolerate far too much and receive far too little in return. My parents, Tony and Audrey, and my in-laws Bob and Verna Boutilier, have provided a steady stream of encouragement and support to see this project through to the end. Thank you for letting my dreams be yours as well.

Map 1   Alberta Oil Sands and Heavy Oil Deposits

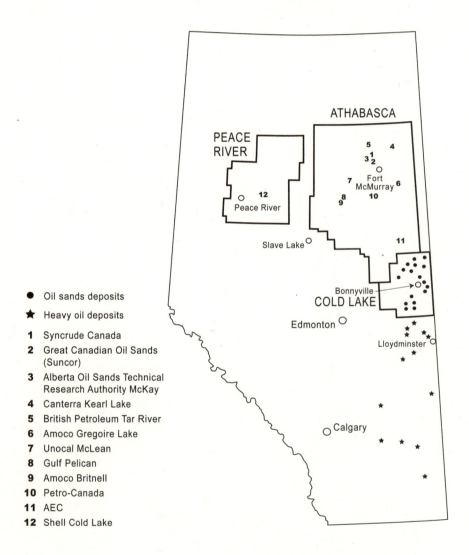

● Oil sands deposits

★ Heavy oil deposits

1 Syncrude Canada
2 Great Canadian Oil Sands (Suncor)
3 Alberta Oil Sands Technical Research Authority McKay
4 Canterra Kearl Lake
5 British Petroleum Tar River
6 Amoco Gregoire Lake
7 Unocal McLean
8 Gulf Pelican
9 Amoco Britnell
10 Petro-Canada
11 AEC
12 Shell Cold Lake

INTRODUCTION

It holds more oil than Iraq's proven reserves of 112 billion barrels. It even holds more oil than Saudi Arabia's 250 billion barrels. In fact, Alberta's oil sands deposit contains between 1.75 and 2.5 *trillion* barrels of oil – approximately 200 billion barrels of which are recoverable with current technology.[1] That is enough oil to supply all of Canada's petroleum needs for the next 475 years. In fact that is enough proven reserves to supply *all* of North America's petroleum needs for the next forty-seven years – without using a single drop of oil from another source. The volume of recoverable oil in the sands is so large that when technological advances prompted the Alberta Energy Utilities Board to include the oil sands as part of Canada's "proven reserves" in 2002, OPEC's share of world petroleum reserves dropped dramatically from 79 to 68 per cent.[2]

The Alberta oil sands, composed of bitumen, rich mineral clays, and water, are not a "new" hydrocarbon source, nor are they a unique resource completely divorced from the overall workings of the Canadian oil industry.[3] Researchers have investigated and debated the oil sands' tremendous potential for North America's modern industrial society over the past eighty years. Canada's early attempts to develop the oil sands gave the country a decided advantage when the development of non-conventional energy sources intensified after world supplies of conventional crude oil began to decline.

To be sure, a few issues still inhibit the oil sands' ability to supplant Saudi Arabia as the world's dominant producer. As a synthetic fuel, the oil sands are capital-intensive and require investors to wait between five and seven years before production begins. After project start-up, access to capital, markets, skilled workers, facility maintenance, and upgrading costs all affect the fiscal bottom line. Over the past two decades, increased diligence by the state and the industry itself to assess and monitor the environmental impact of development also plays an important part in the process. Nevertheless, the growing economic integration of North America, improvements in oil recovery technology and continued instability in the petroleum-rich Middle East present Alberta with a unique opportunity to emerge as a prime, reliable supplier of energy to the twenty-first century industrial world. Industry estimates project that in 2010, the United States will purchase nearly 75 per cent of all oil sands production.[4] With Canada's conventional supplies of crude oil declining at the rate of 500,000 barrels per day, and Canada's status as an important source of crude oil for the United States, any sophisticated discussion of North America's energy resources must assign a prominent role to Alberta's oil sands.[5]

During the twentieth century, interest in the sands waxed and waned with prevailing market conditions as government and the industry alike worked to make the oil sands a viable alternative energy source. This book examines the origins and development of the Alberta oil sands industry over the last century and incorporates findings from several disciplines including history, political science, economics, enhanced oil recovery technology and international relations in an attempt to present the reader with a single-volume history of the development of Alberta's oil sands. The first part focuses on the early history of the oil sands and their development through 1945; the second half discusses the challenges associated with turning a scientific investigation of the oil sands into a commercial endeavour.

By necessity imposed by the long timeframe under discussion, the text takes a narrative form to deal with the issues and difficulties encountered by government and industry in bringing the oil sands to market. But at its heart, the history of oil sands development illustrates the importance of the role of the state in natural resource development, the close ties between the oil industries of Canada and the United States, the pace of technological change and innovation, and the competitive nature of the international oil industry.

The interaction between these four separate entities – the state, the oil industry, the scientific community, and the world petroleum market – created a dynamic and challenging environment capable of astonishing feats of co-operation and commerce. Indeed, when Great Canadian Oil

Sands began production in 1968, it represented a remarkable achievement: a Canadian company, backed by the investment capital of a U.S. multinational corporation, used a separation process researched and developed by scientists funded by the governments of Canada and Alberta to produce a synthetic oil capable of competing against conventional Saudi crude in world oil markets

Careful study of the evolution of the oil sands industry provides a window on the role of the Canadian state in the growth and development of an important natural resource industry. Many years ago, Hugh Aitken noted that the state's traditional role in Canadian development "has been that of facilitating the production and export of ... staple products. This has involved two major functions: planning and to some extent financing the improvement of the internal transport system; and maintaining pressure on other governments to secure more favourable terms for the marketing of Canadian exports."[6]

Certainly, the oil sands story reflects the mixture of public and private enterprise described by Aitken. Crude produced from the oil sands competed against crude from a variety of national and international competitors. As an international price-taker, the oil sands depended upon market prices to guide profitable private sector development. These realities encouraged government involvement in oil sands development from the earliest days of the industry. The active support and co-operation of these two groups – the state and the private sector – spurred scientific research and technological development of the oil sands. Eventually, the public/private partnership enabled the oil sands industry to evolve from a marginal source operating on the periphery to a viable alternative supply.

In nurturing the growth of the oil sands industry, both federal and provincial governments confronted the problem of encouraging development of an expensive commodity while trying to maintain control over the pace of development. Through a variety of organizations, like the Alberta Research Council, Abasand, or the Alberta Energy Corporation, the provincial government in Edmonton shared the risks of development with the private sector. At other times, government remained content with a more limited role to influence oil sands development. Political scientist John Erik Fossum observed that the state can rely on a range of indirect measures to facilitate or hinder progress in energy-related endeavours. Less invasive, but in some instances more effective, indirect measures enabled the state to influence the environment for private sector development and included alterations to taxation regimes, royalty provisions, and establishing the regulatory framework.[7]

With the notable exception of the period during the Second World War, the federal government's energy policies confined themselves to the indirect measures described by Fossum before 1970. After all, Alberta's provincial government received exclusive jurisdiction over the development of natural resources in 1930, except once they passed over provincial or national borders. Furthermore, multinational oil companies and the Canadian industry governed their own relationships. As economic historian Diane Kunz observed, similar oil policies in the United States had the added benefit of being "both financially and politically cheaper" than formalizing relationships with trade agreements.[8]

Ottawa's *laissez-faire* approach deferred to the expertise of the industry and Alberta's pre-eminent position. It also enabled the continental orientation of the Canadian industry to proceed unhindered. Ottawa's benign neglect of the oil industry persisted until the beginnings of the oil shocks when energy issues became highly politicized and prompted a more invasive federal approach into all aspects of the industry. The state adopted many measures, including taxation, regulation, and the creation of a Crown Corporation, Petro-Canada, to foster energy self-sufficiency. When the direct intervention of Pierre Trudeau's government into private sector ultimately failed because of a combination of faulty planning and unfulfilled expectations about world crude prices, the federal government reverted to indirect measures to control development.

State-sponsored research at both the provincial and federal levels between 1920 and the 1950s focused on establishing and perfecting a separation method capable of liberating the oil locked in the sands, and gave birth to a new industry. Later, technological advances and innovation altered industry perceptions of the oil sands by taming production costs and creating new opportunities. Instead of trying to produce large volumes of oil to offset operating costs, continuing research and development enabled the industry to focus on lowering production costs to compete with conventional crude.

The new approach resulted in modernized plant operations with high-tech equipment and computers that also lowered the energy required to produce a single barrel of oil.[9] Current levels of technology and the world price of oil, rather than the level of government subsidies, now determine production levels and make the oil sands a viable alternative to supply North America's voracious appetite for energy.

Given the active participation of U.S. multinationals in the Canadian oil industry, the proximity of the U.S. export market for Canadian crude, and the American influence over the world petroleum market, any discussion of the development of the oil sands must acknowledge the crucial role played by the United States. American multinational corporations and their Canadian subsidiaries provided a much-needed source of investment capital, technology, and expertise – a point generally acknowledged in many studies of the Canadian industry.[10] Furthermore, the large American export market provided the opportunity synthetic producers needed to prove the viability of their product without unduly harming Alberta's conventional oil industry. As the text makes clear, the timing, size, and scope of the first commercial oil sands facility was directly influenced by the demands of the U.S. market. In fact, it is difficult to envision an oil sands development scenario that does not prominently feature U.S. investments, markets, or multinational corporations.

More recently, the terrorist attacks of late 2001 and increasing instability of the Middle East imbued energy issues with a security dimension. Conservation and the use of renewable energy sources (hydro, wind, and solar power) are important and worthwhile initiatives to reduce North American dependence on imported Middle Eastern oil. Nevertheless, the reality is that oil remains one of the most versatile and efficient energy sources for an industrialized economy; globally, oil and petroleum products currently provide between 35 and 40 per cent of the world's energy and the U.S. Department of Energy only expects its market share to grow.[11] Over the next decade, the DOE projects that consumption of oil and petroleum products will *increase* between 1.7 and 2.1 per cent per year to 2020. Meanwhile, the percentage of energy provided by renewable resources plateaued at 8 per cent and the International Energy Agency estimates that worldwide use of renewable sources will drop by 1 per cent over the next twenty-five years.[12]

Other steps to reduce North American dependence on imported oil, including fuel cell technology, remain promising possibilities. However, researchers cannot offer any guarantees that fuel cells will become a feasible alternative within the next generation.[13] Stated simply, oil occupies the dominant role in world energy supplies and it is extremely unlikely that another fuel will challenge petroleum's supremacy anytime soon. The question for concerned politicians, pundits, and citizens alike is where will the next generation of North Americans get the oil to fire the economy, fuel their cars, and heat their homes when energy policy must factor security consideration into the equation? Because of the development of Alberta's oil sands, the Western Hemisphere apparently begins the new millennium with enough non-conventional hydrocarbons, available at relatively cheap prices, to extend the petroleum age for several more generations.

CHAPTER 1

Early History
of Oil Sands in Alberta

At about twenty-four miles from the Fork, are some bitumenous [*sic*] fountains, into which a pole of twenty feet long may be inserted without the least resistance. The bitumen is in a fluid state and when mixed with gum, or the resinous substance collected from the spruce fir, serves to gum the canoes. In its heated state, it emits a smell like that of sea coal. The banks of the river, which are there very elevated, discover [*sic*] veins of the same bitumenous [*sic*] quality. – *Alexander Mackenzie Diary, 1789*[1]

During the first half of the twentieth century, scientists and researchers explored the possibility of extracting usable oil from the sticky confines of the Athabasca tar sands. Generally, a particular crisis, or crisis-like situation, spurred their inquiries. In the aftermath of the First World War, widespread fears of a shortage of conventional crude sources inspired early North American attempts at developing a synthetic fuels industry.[2] Meanwhile, researchers in Alberta attempted to find a practical use for oil sand, particularly as road-top asphalt. Later, as researchers refined their experiments, they focused on developing a process that could separate usable oil from the oil sands. The ultimate goal was to refine the sand and then use it for fuel. For the Province of Alberta, the oil sands represented an opportunity to diversify its economy away from agriculture and enhance prospects for

1

sustainable growth. Alberta's scientists tried to prove that the oil sands were viable as an alternative fuel supply.

Despite repeated assurances that scientists had discovered adequate methods to separate the oil from the sand, it proved difficult to translate success in the laboratory into a thriving commercial venture. Furthermore, a simmering debate between the province and the federal government threatened future use of the oil sands as the two sides battled over natural resource jurisdiction until 1930. Although an agreement eventually granted the province oil and gas rights the issue of oil sands development remained unsettled through the Second World War. The federal government retained possession of large sections of land containing bituminous deposits until after the war.[3]

EARLY FEDERAL EFFORTS

Centuries before the sands were used as a source of fuel, members of Canada's First Nations knew about their existence and used the tarry material to patch canoes. Peter Pond first noted the existence of a black, tar-like substance oozing from the ground in 1778. Other explorers, including Sir Alexander Mackenzie of the North West Company, soon followed and provided the first written description of the deposit. Rather than classifying the sands as a geological formation, Mackenzie believed that the deposit contained "bituminous fountains."[4]

By and large, however, the tar sands remained undeveloped until after the Canadian government purchased the western territories known as Rupert's Land from the Hudson's Bay Company in 1869. Eager to assess the region's potential resources, the Dominion government instructed the Geological Survey to collect samples, make maps, and report on the mineral wealth of the newly acquired territories. Despite an earlier trip to the area by John Macoun in 1872, Robert Bell's 1882 expedition to the Athabasca Basin would be the most important and far-reaching. A geologist by training, Bell examined the oil sands and concluded that the oil sands were part of a vast underground petroleum reservoir. If wells were drilled deep enough, Bell was certain that the result would be a large producing field.[5]

Bell's theory attracted a number of geologists, speculators, engineers, and entrepreneurs to the region for decades. Even the Dominion government got involved and ordered the Geological Survey to drill a well at Athabasca Landing in 1894. Despite reaching a depth of 1,100 feet, the Survey's engineers could not find the petroleum source promised by geologists. The failure of this first well did not dissuade engineers, nor did it discredit Bell's theory. Instead, officials believed they had simply drilled in the wrong location. Three years later, the government tried to drill another well further down the river at Pelican Rapids. This time, although the project produced some natural gas, prospectors were unable to strike oil and the government cancelled the project.[6]

Although direct efforts at tapping the oil sands proved unsuccessful, the Dominion Government remained influential in oil sands development through other means. From 1896 to the outbreak of war in 1914, the Canadian Prairies benefited from one of the largest and most dramatic population migrations in North America. Settlers poured into the region from Ontario and Eastern Canada, the United States, Great Britain, and the European continent. Alberta's population rose from 73,022 in 1901 to 373,943 in 1911. To facilitate the settling of the west, the federal government offered generous terms. Settlers were entitled to a free section of land provided they lived on it and "proved it up" within three years.[7]

Although the Dominion offered settlers these favourable terms, Ottawa remained leery of turning control of the natural resources over to the Prairie Provinces. The provincial government did not possess enough capital to administer the territory of the new province adequately. Political expediency dictated that Edmonton and Ottawa reach a compromise. The federal government agreed to pay the province a special subsidy, based on population, instead of turning full control over the mineral rights to the province. By the terms of this agreement, the Crown retained the mineral rights for approximately 95 per cent of the province, including control over the Athabasca deposit.[8]

Beyond these internal considerations, Canada's oil policy retained international importance because of Canada's status as a British Dominion. Alberta entered confederation (1905) at roughly the same time that London encouraged Ottawa to develop any available oil and gas resources in Dominion territories. Although British oil interests concentrated in the Persian Gulf, the possibility of finding oil within the empire always aroused a certain degree of attention in London. Potential fuel deposits in Alberta piqued the curiosity of both the British and federal governments. British concerns about securing adequate fuel sources found a receptive audience in Ottawa and ultimately encouraged the federal policy of retaining subsurface mineral rights.[9]

The policy gave Ottawa control over large sections of land but did not provide the means to develop the resources they held. The oil sands languished because the federal government did not have the financial resources to develop them. Even conventional oil and gas development stumbled because of a lack of available funds. By the end of the decade, Ottawa effectively conceded defeat and turned the development of the region over to the private sector. The Dominion Land Agent's office granted petroleum and natural gas leases on a first-come, first-served basis. Leases were available to any person who could pay a $5 filing fee and afford rental payments of twenty-five cents per acre for the first year. Thus, for a modest investment of $45, a promoter could lay claim to the mineral rights for 160 acres of land. In 1914 alone, more than five hundred different oil companies began operations and bankers later estimated that unscrupulous promoters raised more than $1 million from private investors.[10] Despite the flurry of activity, fewer than one in ten promoters bothered building a derrick and fewer than that bothered drilling a well.[11]

The situation was no better in Athabasca. Individual entrepreneurs envisioned several ambitious schemes to exploit the oil sands. Some tried using traditional drilling techniques to strike oil, while others attempted to separate the oil from the sands. The reality however, is that very few people possessed enough technical knowledge to make the development of the oil sands even a remote possibility. Typical of these efforts is the story of Count Alfred von Hammerstein, a German emigrant to Canada.

In 1897, von Hammerstein came to Alberta while on his way to the Yukon to take part in the gold rush. Once he arrived in the Athabasca area and heard rumours about a large oil field existing on the outskirts of town, he decided to stay. Rather than searching for gold, von Hammerstein gambled on tapping the huge pools of oil suspected to cause the seeping oil from the sands. In 1910, von Hammerstein notified the federal Department of the Interior that he had struck oil and that he wanted to purchase the land outright from the federal government. Under the regulations of the day, von Hammerstein would have to prove to the Minister that the land, in fact, could produce oil. According to one Fort McMurray resident, von Hammerstein "poured a barrel of oil down a hole, drilled for oil and found it." With the stringent requirements satisfied, the federal government granted von Hammerstein the right to explore 11,405 acres of land.[12]

The haphazard development policy met with growing unease in the province. While the federal government could rightly claim that development of natural resources took place, more socially conscious observers saw some warning flags. Was oil sands development efficient and effective? Most important, where would research money come from? As the number of failed private ventures multiplied, so did the frequency of fraudulent stock promotions that robbed several citizens of their life's savings.

Alberta's first premier, Alexander Rutherford, who occupied the office from 1905 to 1910, believed that the province's economy should develop through co-operation between government and business. The reason seemed logical enough: without a healthy, vibrant business community, building the province's economy would be difficult. "Rutherford," wrote business historian Henry Klassen, "sought to stimulate the growth of Alberta through close ties with business enterprises, believing he was acting in the public interest."[13] Edmonton provided public funds for a variety of projects. The provincial government helped to subsidize the growth and development of several sectors of the economy, including ranching, farming, and manufacturing. Under Rutherford, the province also committed itself to constructing the necessary infrastructure to stimulate business expansion. The creation of Alberta Government Telephones (1906) and Edmonton's interest in expanding transportation links throughout the province established the government's reputation as an active participant in the economy. "Rutherford," concluded Klassen, "assimilated a particular kind of political economy into his vision of Alberta."[14]

Largely due to Rutherford's early effort, a core of community leaders – legislators, businessmen and academics – coalesced during the province's first decade to guide economic development. Soon, the University of Alberta emerged as one key institution for provincial development. Founded in 1908, the university dedicated itself to serving the public interest.[15] Dr. Henry Marshall Tory, the president of the University of Alberta, worried that unscrupulous developers simply wasted the oil sands and fleeced the public. Ottawa's decision to turn control of the development of the resource over to the private sector amounted to little more than policy by neglect. The federal policy stood in stark contrast to Alberta's emerging tradition of public/private co-operation. Anyway, the province had to mitigate the deleterious effects of these "businesses" on the community. Tory believed that the current state of technology would not allow for the successful exploitation of the tar sands. Instead, Alberta should conduct subsequent research and inquiry for the public good.[16]

Transplanting traditional mining techniques from Europe to North America worked for most mineral deposits, but the Athabasca oil sands were a different creature altogether. Oil sand oil is very heavy and has a higher specific gravity than water – meaning that the oil would sink in water rather than float like other oils. Second, tar sand oil does not flow like lubricating oils. It is very viscous at the best of times, prompting one observer to compare extracting oil from the sand to removing honey from a bowl of sugar.[17]

The growing importance of oil for the defence of the Commonwealth prompted the Canadian government to re-evaluate its oil policy. On June 10, 1913, the British Admiralty recommended that the Dominion government set aside certain oil-bearing lands in an emergency reserve. The colonial secretary, Lord Harcourt, reminded the Canadian government that official British policy had changed. The Crown must obtain reliable fuel supplies when they were most needed.[18] As far as possible, the secretary suggested that development of natural resources continue in a manner favourable to British interests. Thus, the British Crown or its representatives (in this case, the Canadian government) should exercise direct control over both the property and the company engaged in exploration.[19]

After receiving instructions from London, the federal government would only issue crown oil leases to Canadian or British companies. In addition, the federal government, under heavy pressure from the British, placed the oil sands in reserve in 1913, effectively ending all speculative ventures. Although Ottawa permitted some private initiatives, their importance was subordinate to the interest of the federal government. As David Breen explains, going beyond the original British suggestion, Ottawa inserted a proviso into the regulations permitting federal authorities to repossess any location under Crown lease at any time. The government could seize any machinery and production facilities on leased land. "The minister of the interior," explained Breen, "clearly intended it to further discount the need to create an Admiralty or government reserve."[20] Thus, until 1930, the province would have to apply to the federal government before conducting research on the tar sands.

Early in 1913, the Edmonton group sent a letter to the federal government inquiring about the current state of federal research on the oil sands. The letter ultimately found its way onto the desk of Sidney C. Ells, a researcher and assistant to the director of the Mines Branch. It did not take Ells long to realize that the Mines Branch possessed very little information about the tar sands. In his memoirs, Ells recalled that a single official report, dated 1883, existed in the government records. Although Ells saw the need to gather more information, he could not convince his superiors at the Mines Branch. The oil sands so enthralled Ells that he almost resigned when his superiors denied his request to study them further.[21] Faced with the prospect of losing a valued researcher, the Mines Branch relented. Ells travelled to Fort McMurray in the summer of 1913 to survey the region and collect samples.

In the year before the outbreak of the First World War, the area around Fort McMurray, except two track surveys, was still unknown and uncharted territory. Fewer than two hundred people lived in the area and the hamlet of Fort McMurray consisted of twelve primitive log buildings haphazardly built around a square. The means of physical communication were as primitive. No railway, no road, no telegraph communications existed and the mail arrived four times a year, conditions permitting. Thus, Ells's trip required extensive preparation and effort to succeed. All of Ells's supplies, including food and equipment, arrived in the camp using the most basic methods – poling, tracking, and portaging. Backbreaking work at the best of times, but in the rugged back country, it was almost too much. Ells asked his crew to track nine tons of equipment and samples to Grand Rapids from Fort McMurray – a distance of eighty-two miles. The trip physically incapacitated three men. Others simply disappeared into the brush.[22]

Meanwhile, at the behest of Dr. Tory, the University of Alberta's chemistry professor Adolph Lehmann began his own research on the bitumen contained in the tar sands. At the centre of his research, Lehmann focused on the question of whether or not the tar sands contained substances similar to those found in European coal tar. If so, this would be a discovery that could herald the emergence of a lucrative chemical industry in Alberta.

Although their respective approaches were different, both Ells and Lehmann sought to unlock the secrets of the physical composition of the tar sands. The two men also devoted considerable time and attention to finding an economical way to liberate the bitumen from the sand. The outbreak of the First World War accelerated both men's research on the oil sands, but for radically different purposes. After August 1914, Lehmann focused on the search for a derivative of the tar sands that might contain substances needed in the manufacture of explosives for the war effort. Ells rejoined his unit in Europe in the spring of 1917 and would remain overseas until 1919. Not wanting to abandon the project entirely, Dr. Tory began to gather all the available research on the oil sands, including copies of Ells's research notes from the Mines Branch. Before sending the material to Dr. Tory, Eugene Haanel, director of the Mines Branch, asked two junior scientists in the department, Dr. Karl A. Clark and Joseph Keele, to assess Ells's writing. More important, Haanel asked the two scientists to decide whether the material was ready for publication.

Clark and Keele were unimpressed with the material. Ells's vast work "had all the appearances of being a hopeless mess" that would require a great deal of effort to organize and synthesize. Nevertheless, Clark and his assistant went through the material and prepared a very critical appraisal of what they read. They could not deny that Ells's research pointed in some very promising directions, but they could not ignore that it did not approach its subject systematically. The muddled and inconclusive draft would require significant rewriting and re-drafting to make the manuscript publishable.[23]

When Ells's material arrived in Edmonton, along with Clark's report, Tory and Lehmann both reached the same conclusion: bringing Ells to work at the University of Alberta would be pointless. His personality was too strong, and his research too different, to mesh with the work being done in Edmonton. The final straw, however, came in 1919 when Ells applied to visit the oil shale deposits near Edinburgh. Tory assumed that Ells was working for the provincial government as well as the federal government and wrote a letter to support his application. As such, Tory believed Ells should provide a written report to the Legislative Assembly to publicize his findings, since provincial taxpayers' money funded his research. Ells was reluctant to provide the report, and his standing in Dr. Tory's eyes dropped precipitously. The president of the University of Alberta decided to continue without the involvement of the federal government.[24]

THE FEDERAL-PROVINCIAL CONFLICT

The end of the First World War marked the beginning of a pronounced political split between the province and the federal government. Both combatants regarded one another with thinly veiled suspicion and contempt. In 1919 wheat farmers – by far the largest voting bloc and wielders of political power in the Prairie Provinces – were in crisis. During the war years, many overproduced or went into debt buying new machinery and land at high interest rates. When the war ended in 1918, those same farmers were left in dire straits. The bottom fell out of commodity prices and farmers expected the federal government to introduce price supports, leading to the brief incarnation of the Canadian Wheat Board.[25]

The Unionist government of Robert Borden desperately tried to shore up its standing throughout Western Canada where support was shaky. Most farmers reluctantly left the Liberal Party and cast their ballots for the Unionist government because they believed Borden's pledges to reform the Canadian government. By 1919, however, with fewer reform initiatives coming from the conservative government, most voters abandoned the Borden coalition and created a new political party. After many lengthy trips overseas, Canada's wartime leader was tired and ill. Borden stepped down as party leader in July 1920, taking with him the moniker of the Union Party. Arthur Meighen became the leader of the newly named National Liberal and Conservative Party and promptly began to alienate many voters. Meighen regarded the Prairie Provinces with thinly veiled scorn, greeting agrarian protests with the haughty attitude that the country's western farmers were "Socialistic, Bolshevistic and [full of] Soviet nonsense."[26]

While the federal government reacted hesitantly to demands of farmers, the political context continued to sour. The end of the First World War did little to alter the *status quo* on natural resource jurisdiction; Ottawa retained possession of the resources in Western Canada and cries from the provinces grew louder. If anything, positions on both sides began to harden, only now the debate attracted more attention. The Prairie Provinces maintained that, since they did control the territory under their jurisdiction, the federal government should compensate them for the alienation of resources that had taken place.[27]

However, Ontario and Quebec argued that the Dominion of Canada purchased the territory from the Hudson's Bay Company. Ottawa administered the lands because the provinces were unable to do so. If any group was entitled to some form of compensation, it was the people of Canada, and not the provinces. The people of Canada purchased the lands, argued J.B.M. Baxter of New Brunswick. "It is impossible to consider Canada as a sort of partnership in which the partners should ask for recognition of their individual interests."[28]

The battle for control over petroleum policy and the future of oil sands research was subsumed within the context of the deteriorating political climate between Alberta and Ottawa. The nation's petroleum policy since 1905 evolved from British concerns about securing usable supplies within the Empire and having resources developed by "reliable" Canadian or British producers.[29] Despite the preferential treatment given to British and Canadian capital, however, the British hardly invested in Canadian industry generally, and in the oil industry in particular.

As late as 1914, British investment accounted for 70 per cent of total foreign funds in the country. Within four years of the end of the conflict, however, the United States assumed the role of Canada's single-largest investor and never relinquished the title.[30] In that time, the tangible manifestations of American financial wealth became increasingly apparent. By 1920, an American firm, Jersey Standard, came to dominate Canada's petroleum industry through its Canadian subsidiary, Imperial Oil. Imperial Oil managed to circumvent federal regulations, which required oil companies to have a majority ownership in either Canada or the United Kingdom. The company bought freehold leases from private mineral owners – mainly the Canadian Pacific Railway and the Hudson's Bay Company.[31]

The question in Ottawa was whether Canada could afford to continue its preferential system or whether the government should loosen its restrictive practices and allow American capital into Canada. The matter gained further urgency when the United States challenged the San Remo agreement (April 1920) to divide Middle Eastern oil rights. British Prime Minister David Lloyd George and France's Alexandre Millerand decided France would get 25 per cent of the oil from Mesopotamia and would assume the German share in the region. In return, the French gave up their claims to Mosul. Lloyd George made it clear that any development of Mesopotamia's fields would be done under British control.[32]

The arrangement was an explicit rejection of America's pursuit of the Open Door for world crude. When the British civil commissioner detained two Standard Oil geologists and turned them over to the Baghdad police, the incident caused an uproar in the United States. Americans immediately denounced the San Remo Agreement as high-handed and imperialistic. Congress retaliated by passing the Mineral Leasing Act. The legislation denied drilling rights on public lands to foreign-owned corporations whose governments denied similar access to American corporations.[33]

Confronted with such a clear choice, to support the British, who did not invest in Canadian oil, or accede to American capital and "open the door," the government waffled. It scrapped its controversial ownership requirement on lands farther to the south – specifically the Turner Valley area near Calgary where Standard Oil already had producing fields. But it also reinforced the restrictive regulations in the North West Territories, gambling that the region would yield a large producing field.[34]

While the federal government moved to place the resources of the North West Territories firmly under Ottawa's control, the critical issue in the province concerned the oil sands. Which group – the province or the federal government – would take the initiative in realizing the potential of the oil sands? The federal Honorary Advisory Council entertained very definite ideas about how to proceed on the development of the Alberta tar sands. Moreover, the Council knew whom they wanted to lead the project – Sidney Ells. Dr. Tory, on the other hand, wanted to keep development of the sands in the University of Alberta's laboratories and under Edmonton's control.[35]

Beyond the political divisions and passions the debate inspired were the technical differences between the separation processes advocated by the University of Alberta and the Mines Branch. Lehmann's initial experiments led him to pursue a process that used chemical emulsifiers to separate the oil from the sand, while Ells focused on using superheated steam. Although both processes displayed considerable potential, subtle signs emerged to suggest that the Honorary Advisory Council had already decided which direction the research should go.

In 1920, the Mines Branch endorsed the findings of two scientists at McGill University who announced that oil sand samples contained "ingredients belonging to the naphthalene series." The announcement essentially declared that the oil sands were suitable only for producing road asphalts and not petroleum oils. Several scientists at the University of Alberta simply did not accept the findings of the McGill scientists. Their research led them to conclude that the oil sands had application well beyond that of road asphalt and Tory moved to resolve the issue quickly. In a letter to the Mines Branch, Tory pointed out that the composition of the oil sands did not justify the McGill findings. The oil sands contained the same asphaltic hydrocarbons contained in California heavy oils and not the paraffin hydrocarbons common in Pennsylvania crude, with which the scientists were more familiar. Tory suggested that the Mines Branch should not reach any hasty conclusions about the oil sands with incomplete information.[36]

Despite Tory's protests, the government stood beside the McGill findings. Dr. Lehmann soon found that the federal agencies were reluctant to grant him additional research money to investigate the potential of chemical emulsification techniques. The sudden reassignment of one of Lehmann's protégées created suspicions about the direction of federal research.[37] Coincidence or not, the decision was final, and the province decided to forge ahead alone. The province, in conjunction with the University of Alberta, put the finishing touches on its newly created Scientific and Industrial

Research Council – the precursor to the present-day Alberta Research Council. They ordered the organization to investigate the properties of the oil sands systematically and to develop a suitable separation process.

The Scientific and Industrial Research Council of Alberta was the first provincial research organization of its kind in Canada. Initial projects focused on using the hot-water separation method, but in the spring of 1920, Tory received word that Dr. Clark – Ells's erstwhile critic in the Mines Branch – had successfully separated oil from the tar sands. Clark's initial findings and scientific approach appeared promising to Dr. Tory, and the latter invited Clark to visit him at the University of Alberta. An introductory meeting between the two men went very well and led to an exchange of letters before Dr. Tory offered Dr. Clark the position of Research Professor. Although many others, including Dr. Lehmann and Sidney Ells, had researched the tar sands before, Dr. Tory instructed Clark to examine the problem with "fresh eyes." Dr. Clark thus began his career with a clean slate and started his research from scratch.

Clark's appointment also guaranteed that federal and provincial research efforts would remain divided. The clash of personalities between Ells and Clark personified the split between the province and the federal government. Ells believed that he had several "enemies" in the scientific community bent on discrediting him. He told the Alberta premier that Eugene Haanel opposed him because Ells unmasked his pro-German sympathies during the First World War.[38]

As for Clark, Ells believed that the Alberta Research Council's chosen oil sands expert selfishly wanted to usurp Ells as the pre-eminent researcher in the field. According to Ells, Clark's work so closely paralleled his own that Clark needed to isolate Ells from other members of the scientific community. Ells never seems to have accepted that his work, although original and important, was sloppy. Even as Ells continued to guide federal efforts, there was growing unease about his methodological practices in the scientific community that detracted from the substance of his research. Thus the federal effort, so vibrant between 1913 and 1917, began to stumble in the inter-war years as the initiative passed to the Alberta Research Council.[39]

There may have been some merit in Ells's claim that Clark was simply copying his work. Although his research began with "fresh eyes," Dr. Clark provided few new findings. The value of Clark's contribution to the oil sands effort, however, is beyond question. He systematically investigated many claims made by Ells and remained instrumental in developing and promoting a commercial extraction process. He confirmed that the sands were the physical product that remained after lighter substances – natural gas, for example – escaped, leaving behind natural asphalt. Most important, Clark determined that the tar sands were not a direct source for gasoline or kerosene given the state of refining techniques in the 1920s. Through a process known as "destructive distillation," tar sand did yield some petroleum products.[40]

Scientists could only recover 45 per cent of the material by weight and would lose the remaining 55 per cent, making destructive distillation a very wasteful process. Perhaps, if given enough time and patience, research and development into new areas might change those results. Researchers first tried different types of chemical solvents to extract the bitumen from the tar sands – a method first tried in the Western United States. However, not only was the cost of the solvent high, but its loss in sand tailings discouraged further development.

HOT-WATER EXTRACTION

Stymied in these two approaches to separation, Clark turned his attention to a hot-water extraction process that displayed a marked degree of success. This third extraction method relied on the use of hot water to separate the bitumen from the sand. Between 1922 and 1929, Dr. Clark gradually developed a rudimentary process to separate the oil from the sands. In the basement of the University of Alberta's powerhouse, Dr. Clark and Sidney M. Blair built their first hot-water separation plant in 1923. They fed mined sands into rotating conditioning drums – very much like the large drums used to mix cement. Once in the drums, a combination of hot water and steam were tumbled with the sands to produce a pulpy matter. Before pumping the matter into separation cells, screening removed lumps and other debris. The bitumen would then separate from the sand and water because of a difference in density. Both water and sand are much denser than bitumen, allowing researchers to skim the bitumen off the surface of the sand-water mixture where it floated. However, the separation was still not complete. The oil froth that resulted from the first round still contained fine solid materials and water that need filtering. The sand and water material from the initial separation underwent the same procedure a second time because it still contained approximately 20 per cent of the original bitumen content.

Perhaps a little overconfidently, Clark told Dr. Tory in late 1921 that hot water separation provided the solution the Research Council was looking for. The most important job for the Research Council and the province now would be the public education of Canadian citizens about the potential of this resource. Still, Clark was willing to concede that a few more practical engineering questions remained to iron out. The most important was "the practical application of the new method to the production of bitumen from the tar sands."[41]

Nevertheless, Clark was confident that the hot water separation process was more than adequate to the task. The province should now use the sands as a road-top asphalt. "Public opinion," wrote Clark, "uninformed about all the factors involved, impatient for tangible results and with preconceived ideas on the subject of tar sand exploitation, would probably call for a quick 'topping off' of the investigation in some ostentatious way."[42] Clearly, it was up to the province and the Research Council, in particular, to serve as a steadying influence to ensure that they did not waste the oil sands.[43]

It is surprising that the Research Council focused on the oil sands' use as a paving material and not as an alternate fuel source. This is more remarkable upon consideration of contemporary concern about the dependability of North American oil supplies. In 1918, for example, Canada's annual domestic consumption of oil was more than eleven million barrels while the country's annual domestic production amounted to only 250,000 barrels. The gap between supply and demand was immense because by 1921, domestic production met less than 1 per cent of demand.[44]

Similar conditions prevailed across much of the rest of the continent as soldiers returned home from war and the desire for motor vehicles increased. In the United States, between 1914 and 1919, the number of registered motor vehicles jumped from 1.8 to 9.2 million.[45] Meanwhile, the director of the United States Bureau of mines predicted in 1919 that within the next five years, production from U.S. oil fields would plateau. America, he concluded, would face an ever-increasing decline in oil reserves thereafter. In 1920, the Pacific Coast experienced a severe oil shortage, much to the alarm of the local inhabitants. Only a temporary restraining order stopped Commander John Mel from landing marines in California and seizing oil from the General Petroleum Company.[46]

Gradually, as a response to the potential threat of an oil shortage, the United States, Britain, and Canada would cautiously begin developing synthetic fuels programs to fill the need. Rather than rely on a single-track method to meet the challenge, both the British and the Americans expanded outward and looked to discover other producing fields. Thus, as William Beaver points out, when oil was scarce in the early 1920s, the initial response of government was not to launch expensive synthetic fuel projects. Instead, government helped look for alternative sources of conventional crude. Increased exploration in Latin America and the Middle East "reduced any sense of urgency to explore synthetic fuels," he observed. *"As long as foreign oil could be obtained at reasonable prices, the difficult task of developing synthetics could be averted."* [47]

Several changes to conventional production methods influenced the pace and development of synthetic fuels. The 1920s witnessed many fundamental changes to the North American petroleum industry that helped to increase production and lower crude prices. Rising world prices for crude oil in the early 1920s provided the necessary stimulus for further exploration and led to large discoveries in Oklahoma, West Texas, Louisiana, California, and New Mexico. Meanwhile, oil companies adapted technology developed for the First World War to aid exploration. Combined with advancements in refining and cracking, a veritable flood of oil dispelled the gloomy predictions of the early 1920s.[48]

Lacking the momentum to sustain interest and investment, synthetic fuel projects languished in corporate boardrooms and, to some extent, as a public policy issue. Partly as a result, Clark hesitated to publish his more optimistic predictions in Research Council publications through the early and mid-1920s. Perhaps Clark recognized that his earlier claims to have "solved" the separation problem were too optimistic. Until 1927, Clark's published research focused on the value of the oil sands as a source of asphalt. He argued instead that the sands were unlikely to be amenable to petroleum cracking. Gradually, however, opinions in the field were changing and this helped spur development. In 1926, Sidney Ells concluded that the oil sands demonstrated considerable potential as a source of "liquid hydrocarbons." Ells specifically discussed the possibility of converting the processed oil sand into gasoline.[49]

Resounding indifference greeted Ells's findings across North America, except among the small circle of oil sands researchers and the Province of Alberta. The oil crisis of the early 1920s passed, as conventional crude gushed from the ground in increasing abundance. From a business point of view, adding another source of fuel – especially a costly, capital-intensive synthetic fuel – to an already crowded market made little sense. While on a tour of North American refineries in Toronto and Chicago during 1926, Clark found little interest among refiners for producing fuels from the bitumen derived from the oil sands. Nevertheless, the oilmen continued to encourage Clark's research efforts. Although the province continued to sponsor research into the field, without a viable commercial path for the research to follow, interest in the project waned. Ultimately Clark feared that the momentum needed to sustain his research had turned into inertia. The problem facing the project was clear. How would research go on even if government gave the signal? Every researcher previously connected with the oil sands was gone.[50]

Matters would soon change for the better. In fact, the decision to continue with development and research made a great deal of sense. In 1926, Canadian crude production rose slightly and now accounted for approximately 1.6 per cent of all the oil consumed in the country. Clearly, with imports meeting more than 98 per cent of fuel consumption, Ottawa would welcome any opportunity for the country to increase its domestic production. Moreover, as Clark explained, Western Canada did not generally benefit from lower world prices for imported oil. Although the price of a barrel of crude oil on the world market fetched around $1.00, factoring in transportation costs meant that Albertans were paying approximately $4.00 per barrel. It only made sense to begin developing the province's own resources, provided it was economically feasible.[51]

Other pragmatic reasons existed for Edmonton to continue pressing for further development. Several years before, the Research Council's optimistic reports encouraged provincial policymakers about the future of oil sands development. Believing that a solution was close at hand, Edmonton committed itself to the sands' development with an outlay of cash to complete a railroad from Edmonton to Fort McMurray. Five years later, the province had little return to show for its investment. If, however, a practical commercial application for the oil sands was near, the province would have an opportunity to gain a return on its investment. In addition, Ottawa and Edmonton finally addressed lingering questions about natural resource jurisdiction with an eye toward finding a solution.

The province was the beneficiary of a decade-long attempt by the federal government to restore national unity in the aftermath of the First World War. Early in the decade, the Prairie Provinces and the federal government began negotiations to turn control of natural resources over to the provinces. Even if discussions had been long and protracted, and sometimes acrimonious, the province and the federal government now appeared to be close to resolving the issue.[52]

As a demonstration of good faith, both the province and Ottawa agreed to develop the Athabasca tar sands jointly. In May 1929, the Alberta Research Council and the Mines Department of Canada signed a two-year agreement to form the "Bituminous Sands Advisory Committee." The newly formed committee would finance and administer two years of research. Alberta would work on the construction and operation of a test facility while the federal group would conduct further research on the raw material itself. Finally, after years of dividing their efforts, Edmonton and Ottawa were working together.[53]

On the surface, the new initiative was a godsend for Clark. Years earlier, Clark questioned the wisdom of allowing the province to continue independently of the federal government. In one letter to Blair in 1927, he noted with some frustration that three different researchers were examining the same substance with little interaction or coordination between them. "That is the show that is supposed to evolve the solution for a big railway problem of a province," wrote Clark. Perhaps more frustrating to Clark was the fact that "not one of the three workers has enough backing to accomplish anything."[54] Now, the situation was radically different. Instead of living hand to mouth, the Alberta Research Council had appropriated $30,000 – spread over two years – to allow researchers to prepare and plan a comprehensive program. Moreover, because the Mines Branch and the province divided responsibilities between themselves, Clark could devote more of his time to solving problems related to extraction.[55]

The high hopes soon dissolved. Clark expected clear lines of demarcation between provincial research and the Mines Branch, but none existed. In May 1929, Clark learned that Sidney Ells continued with plans to build another separation plant independent of the Research Council. This new plant, funded by Ottawa for $50,000, would test a different hot water separation process from the one designed by Clark.[56] The differences were enough for two separate patents even if the two processes were strikingly similar.

Some troubling questions arose from this turn of events. Did Ottawa intend to pursue its own path on oil sands research? When questioned about the move, the federal government blandly replied that several other provinces were interested in the separation methods under investigation. Ottawa believed they would get the information more quickly with the Dominion directing the project.[57] With its fears allayed, the province let the matter drop. Blocking Ells could gain nothing. "It looks to me as though he has seen through going ahead at his own place," wrote Clark in his field journal. "If he were shunted off onto something else, everything would be stalled indefinitely."[58]

Although he was uncomfortable with the government's decision, Clark realized the challenges still inherent in his own separation process; specifically the problem of separated bitumen breaking down because its water content was too high. If he could not solve the dewatering program soon, it could jeopardize the Research Council's progress.[59] Clearly, the plant needed to undergo serious renovations and the question now became how best to implement these changes. Clark believed it would be easier to incorporate improvements into a new plant than it would be to renovate an old one. He thus proposed that the province dismantle the current facility and reassemble it closer to the deposits on the Clearwater River.

When the plant opened on October 29, 1929, it only operated for three days and produced eleven barrels of bitumen. While the bitumen was of high quality – less than 10 per cent mineral contents – its water composition remained high at 30 per cent. The latter figure was 10 per cent higher than any Clark obtained in laboratory experiments.[60] Evidently, the Research Council's emulsifying agent was less than effective.

In late 1929, research on the hot water separation process was clearly entering a critical phase. After a long winter of perfecting their process in the laboratory, Research Council scientists went back into the field in the spring of 1930 to try again. Again, the plant failed to operate smoothly. Clark assumed that they would have an easy time of separating the sand mined right from the quarry outside the facility. He was patently wrong. "Material we had brought down to the lab from there worked nicely," remembered Clark. The same was not true for the sand obtained from around the site. "First we ran into hard stuff with a stiff asphalt in it that would not stick to the skimming wheel. Moreover, tar sand that looked quite all right failed to perform right. Our settling tanks and evaporator for cleaning up the crude tar worked fairly well until the evaporator blew up."[61] Beyond the problems encountered with the physical material, the plant itself displayed a frustrating inconsistency. On some days, the process wasted steam and on others, the separator failed to work properly. Still, enough reasons remained for Clark to be optimistic about the plant's future. The process itself proved effective, even if now most of the problems seemed mechanical in nature.[62]

For a moment, it seemed that the province finally turned the corner on oil sands development. The Research Council's positive tests seemed fortuitous; the protracted negotiations between Ottawa and Edmonton about the transfer of mineral rights from the federal to provincial government would come into force. In the summer of 1930, the province and the federal government ratified an agreement that ensured the formal transfer of natural resources would take place on October 1, 1930. Little did Edmonton know that Ottawa was not prepared to cede control of the Athabasca deposit to the Province.

In the fall of 1929, Dr. Clark received a letter from Max Ball, a petroleum engineer from Denver, Colorado. Ball said he was researching the oil sands for clients interested in processing the separated bitumen. Early in 1930, Ball met with the Minister of Mines and the Director of National Parks about the business possibilities. Then, just as the pilot plant at Clearwater was gearing up for its new season, Ball and some of his associates arrived in Edmonton to speak with Clark. From the outset, Ball made it clear that his group intended to apply for a lease along the Horse River Valley and 30,000 other acres along the Athabasca River. In total, Ball proposed to spend at least $150,000 on the development of a commercial plant to separate the oil sands.[63]

Although Ball impressed Clark – "by far the most satisfactory man to talk to that has yet come along with a bituminous sand development scheme" – storm clouds gathered on the political horizon.[64] Ottawa inserted a caveat into the Natural Resources Transfer Act (1930) that granted the federal government control over approximately 2,000 square miles of territory in Northern Alberta, including the Athabasca deposit. Then, a matter of weeks before the agreement was to take effect, Ottawa signed a lease agreement with Max Ball that would allow Abasand Oils to develop the same area that the federal government had held back from the province in its agreement.[65]

Howls of protest rang throughout Alberta when the details became public. Not only did Ottawa sign over mineral rights to one of the most promising oil sands deposits in the province, it did so without consulting the province. The move prompted many Albertans to question whether or not the federal government negotiated the Resource Transfer Agreement in good faith. Public displeasure increased after the Dominion announced that Ball could develop an additional six square miles of acreage anywhere in the Athabasca deposit.[66] Concerned Albertans voiced their darkest fear: the federal government intended to develop the oil sands without the involvement of the provincial government.[67] The Athabasca oil sands were about to become a commercial endeavour through Max Ball's Abasand Oils, Limited.

The remainder of the decade was less than satisfying for the Province of Alberta. The Depression hit the province hard; economic hardship combined with years of drought to cripple the economy. Not surprisingly, with thousands of people struggling simply to survive, the Alberta government decided to disband the Alberta Research Council in 1932 to conserve scarce resources. Progress on Abasand also suffered because investment capital simply disappeared. Developers displayed little interest in synthetic crude oil when wildcatters and overproduction in the United States depressed world crude prices. Although the Depression severely constrained Abasand's operating budget, the company continued to operate for the remainder of the decade and significantly refined its operations. Despite the myriad economic and market conditions during the 1930s that seemed to doom Abasand to failure, the company managed to stay afloat. Abasand continued to sustain the interest of businessmen and engineers in the oil sands.[68]

In retrospect, a fair degree of uncertainty and muddling about distinguished the first decades of oil sands development. Since neither the province nor the country at large could boast of a significant oil industry, the oil sands represented an attractive alternative to costly imports. When world crude prices were high, research and development efforts intensified. When world crude prices plunged, the sense of urgency waned. For a nascent industry that required a high degree of government involvement and capital, fluctuations in funding and interest were difficult obstacles to overcome. Nevertheless, government-sponsored research clearly found a receptive audience as early commercial projects simply expanded on paths of inquiry blazed by government researchers.

The work of the Research Council encouraged the development of commercial enterprises and applications. Although few analysts would argue that Clark's work on the hot water separation process was groundbreaking, all agree that he deserves credit for refining the process and displaying its utility. The close, and at times direct, partnership between both levels of government and business typified traditional government/business relationships of the period in Canada. The state's interest in the development of the oil sands was twofold. First, it conformed to the longstanding policy of directing and aiding western economic development. Second, because of the country's lack of a significant domestic industry, the oil sands represented an opportunity to address the energy issue. Ironically, as David Breen pointed out, even as Canada looked to assert its independence from Britain, British concerns and interests guided the Canadian government's resource policies.

Much to Clark's consternation, the development of a synthetic fuel industry would not provide immediate answers to Canada's growing energy dilemma. Both levels of government began the process of research and development with an eye toward long-term development. The range of problems and obstacles encountered during the industry's first decade – the simmering tension between the province and the federal government, problems with extraction methods, attracting investment capital – foreshadowed many problems the industry would later encounter. Still, researchers made enough progress on the oil sands through the end of the 1930s that elected officials considered the oil sands a feasible source of energy during the Second World War.

CHAPTER 2

Abasand and the
Federal-Provincial Conflict

I can truly state that the government money has not been expended for the purpose it was designed for, that the tar sands did not get any chance and that an additional $350,000 will probably be frittered away in a similar manner unless Hercules and his broom puts in an appearance, and that soon, and he would need a double-action broom that sweeps on top as well as on the bottom. – *Paul Schmidt, Mayor, Fort McMurray, December 1943.*[1]

If all we can get for $1,700,000 is a hole in [the] ground, then it is time we knocked the tar out of someone. – *Honourable W. A. Fallow, Alberta Minister of Public Works, March 13, 1944.*[2]

The Second World War brought tremendous political change to the country as the dynamic of total war altered the landscape. Shortly after Canada declared war in September 1939, the federal government in Ottawa became the most powerful government in the country. Entirely responsible for Canada's war effort, Ottawa took the authority to tax away from the provinces. In exchange, the federal government provided direct support payments to the provinces, reducing the legislative role of the provinces to the maintenance of essential services. "No province during the war years embarked on any new program," wrote Alberta historian T.C. Byrne.[3]

If the federal government was the most powerful political entity, the Department of Munitions and Supply became the most important government agency for Canadian industry. As historians Robert Bothwell and William Kilbourn later observed, the most powerful man was whomever the Prime Minister selected to run Munitions and Supply.

> The Minister could buy and repair, mobilize and construct, requisition and order anything he felt necessary to the production of munitions. There was no barrier to his powers, not even a constitutional one, for in wartime the provinces conceded jurisdiction over provincial resources to the central government. Canada passed in the twinkling of a pen from a free enterprise system regulated by ten jealously competing sovereignties to a centrally directed economy regulated by the government's perception of the needs of the war.[4]

Presiding over this massive undertaking was one of the most competent and driven ministers in Canadian history, Clarence Decatur (C.D.) Howe. Born in Waltham, Massachusetts in 1886, Howe was the most successful businessman/politician of his day. An engineer by trade, Howe built his fortune building grain elevators across Canada and Argentina. However, when the Depression ended his business, he entered the House of Commons as a Liberal in 1935 and joined Mackenzie King's cabinet in 1936.

Howe cut an impressive figure in the House of Commons, having served in several different, and equally important, cabinet positions. He also served as an important link between the Liberal Party and Canadian industry. Perhaps more than anything else, however, Howe earned the respect and admiration of Opposition members for his achievements during the war. Howe's management of the department of Munitions and Supply completely restructured the Canadian economy during the Second World War. With increased federal spending, accelerated depreciation programs, and the creation of some crown corporations, capital investment increased in all sectors of the economy. Meanwhile, the wartime emergency spurred phenomenal growth in the natural resource sector as Canada sought to replace supplies of raw materials captured by the enemy. In all, it was a remarkable accomplishment. "Before the war," reflected Howe later, "few thought of Canada as an industrial power and many doubted whether the manufacturing techniques, which had been perfected in other countries over the centuries, could be developed here in time to produce the war supplies required. It was a challenge that was splendidly met by the men and women of Canada."[5]

While he earned the respect of his contemporaries with his accomplishments, many were less enthusiastic about the way he achieved his goals; several critics charged that Howe had little regard for the role of Parliament. One of Howe's most persistent antagonists, John G. Diefenbaker, observed that Howe thought "the Cabinet was a kind of board of directors; the shareholders had the right to know something of what was going on but had no inherent right to examine carefully or question the policies."[6] Even Prime Minister Mackenzie King wondered if, perhaps, Howe was too power-hungry. Nevertheless, Howe's record of accomplishment spoke for itself: if Parliament wanted tangible results, he could get them.

OIL AND THE WAR EFFORT

Oil alone did not determine the outcome of the Second World War, but naming another resource as vital to the conduct of modern warfare is difficult. Consider that by 1939, most of the American Navy's ships, and approximately 85 per cent of the world's merchant fleets, used petroleum-based fuels to power their engines. During the war, an average armoured battalion required approximately 17,000 gallons of gasoline to travel a hundred miles. During the peak of fighting in the European theatre, the U.S. Army Air Force alone consumed fourteen times the total amount of gasoline shipped to Europe between 1914 and 1918. Between December 1941 and August 1945, the American oil industry produced and processed more than six billion barrels of oil for both civilian and military use.[7]

Since 1930, Canada's domestic crude production rose from 3 per cent of national consumption to slightly more than 18 per cent in 1941. While these strides were impressive, they were clearly not enough to keep pace with the demands of mobilization.[8] Quite simply, the domestic crude situation looked bleak and on November 7, 1941, C.D. Howe rose in the House of Commons and alerted the country to the emergency. Canadian demand for crude oil and finished products for the coming year would be 64,000,000 barrels, an increase of 10,000,000 over the previous year. Turning to questions of supply, Howe could only speculate. He allowed that estimating Canadian production with any degree of certainty would be difficult, but doubted that it would exceed production for the previous year. Increased imports met the projected gap of 54.5 million barrels between available supply and demand. "We must bear in mind," added Howe, "that an oil well is a depleting asset and that we must have new wells and we must find a new Turner Valley."[9]

The urgent need to develop a domestic source of crude oil reflected more than convenience. Canada could not continue to rely on imports to meet demands. For one thing, the toll German U-boats extracted on Atlantic convoys meant that planners directed all available tanker space toward supplying the British Isles.[10] More important, Canada needed to conserve its available reserves of American dollars. Before the outbreak of the war, Canada's trade with its two major partners, Britain and the United States, respectively, remained in relative balance. Canada ran a surplus with Great Britain and a deficit with the United States. Generally, the two totals cancelled each other out because Canada could freely convert pound sterling into dollars. The war radically changed this dynamic. The British forbade the conversion of sterling, already a weak currency, into dollars for the duration, essentially freezing Canadian assets in London and preventing Canada from balancing its accounts. The dilemma grew for Ottawa before the passage of the Lend-Lease Act when Canada served as an important link in the supply route between the United States and Great Britain. Mackenzie King's government increased imports, diminishing the country's precious dollar reserves, and delivered the material across the Atlantic to aid the war effort.[11]

Although Canada could qualify for Lend-Lease, the Prime Minister and his Cabinet decided against applying for aid. The Act's provisions would first require Canada to liquidate all its assets in the United States. Eventually the Hyde Park Agreement of April 1941 would balance Canada's account with the United States, but problems remained. Canada accumulated unconvertible balances of British sterling in London and had to mind its current account deficit accordingly. To staunch the flow of dollars out of the country, the Oil Controller entered negotiations with Socony-Vacuum Oil Company. The government pleaded with Socony to allow Imperial Oil Limited (Standard Oil of New Jersey's Canadian affiliate) to manufacture Socony's patented Mobiloil in Canada. The government also began negotiations with the Ethyl Corporation in July 1940 to find more suitable payment arrangements. Eventually, the discussions resulted in the company taking its payment half in Canadian dollars and half in U.S. dollars.[12]

Before the attack on Pearl Harbor, the United States had extensive contact with the Oil Controller's Office. The early relationship helped to lay "the foundation for close and cordial relations between the two countries with respect to oil and problems of mutual concern."[13] When the Japanese attack on Pearl Harbor brought the United States into the Second World War, contact between the two nations increased and emphasized the "continental" approach to oil issues. The authors of the official history of the Petroleum Administration for War wrote that Ottawa and Washington closely coordinated their oil policies.

> Co-operation between the two countries made possible the fullest utilization of joint resources in meeting wartime demands. The line between Canada and the United States was, practically speaking, to a great extent ignored in order that necessary programs might be carried forward with a minimum of dislocation and inefficiency.[14]

Beyond the close coordination, with the United States as a belligerent, the dollar shortage would ease considerably. In fact, by war's end, Canada enjoyed a trade surplus with the United States – but new problems appeared to replace old concerns. First, Canadian officials realized that mobilization efforts south of the border would jeopardize access to American crude supplies. Beyond shortages induced by the stresses and strains of industrial mobilization in the United States, serious questions emerged about the ability of American reserves to meet demand. While the American oil industry grew significantly in the 1920s and the 1930s, it appeared to slip between 1941 and 1945.

The discovery of new fields dropped sharply during the war and helped to create and sustain a growing pessimism in the oil industry.[15] In 1940, the United States produced approximately 3.7 million barrels per day; that was, roughly, 22 per cent below the estimated productive capacity of 4.75 million barrels per day. Also in 1940, the industry estimated that eighteen billion barrels of oil remained in proven reserves, establishing an ample margin for error. However, after Pearl Harbor, the feeling of complacency dissolved in response to transportation problems, shortages of strategic material, and a sharp decline in drilling activity. Rumours persisted in the oil industry that future discoveries would be more difficult and expensive to bring on stream. In December 1943, Harold Ickes surveyed the state of the American oil industry and concluded that, in the event of another world war, another country would have to supply the oil. "America's crown, symbolizing supremacy as the oil empire of the world," wrote Ickes, "is sliding down over one eye."[16]

Back in Canada, the crude oil situation grew more uncertain through the early months of 1942. The torpedoing of two additional Canadian tankers meant that, as of February 4, Canada's tanker deficiency jumped to 8.6 units, or 8,600,000 barrels of oil. Further exacerbating the situation was a marked decline in crude production from the Turner Valley area of Alberta through the first quarter of 1942.[17] In assessing this information, the federal Oil Controller, George Cottrelle, reported that the oil sands held tremendous potential for the war effort. Cottrelle, a former banker from Toronto and the chair of Union Gas, exercised enormous power and influence as Oil Controller. He could "take possession of and utilize any land, plant, refinery, storage tank, factory or building, used or capable of being used for mining, drilling for, producing, processing, refining or storing of oil," fix sale prices for oil and prohibit any person or business from dealing in oil.[18] Now, Abasand's operations caught his eye.

THE FEDERAL GOVERNMENT AND ABASAND

In 1941, Abasand processed 19,000 tons of sand, producing 17,000 barrels of bitumen converted into gasoline, diesel fuel, coke, and fuel oil. While these figures represented a great success for Abasand, the federal oil controller wanted to know if the plant could process 10,000 tons of sand *per day*. Cottrelle acknowledged that a few important questions remained before the project could succeed. The physical location of the plant was of concern. Abasand also lacked a commercial separation process. Cottrelle argued that Abasand could not continue operations with its present facilities. In fact, it was debatable if the current facility could produce enough material to make a systematic evaluation possible. Regardless, the Oil Controller suggested that the entire operation would eventually have to move thirty miles to the south. In the Mildred-Ruth Lakes area, Abasand would build a new $10 million, 10,000-barrel per day plant. Nevertheless, Cottrelle advised Howe that the government should continue, if cautiously, with the development of the oil sands. "If anyone is likely to solve the problems attendant on this deposit," wrote Cottrelle, "Mr. Max Ball and his associates will do it."[19]

Cottrelle's ringing endorsement of Max Ball and Abasand smacks more of desperation, heightened by Canada's precarious crude oil situation, than it does of abiding faith in the plant. In fact, Cottrelle ignored a steady stream of negative reports to the Oil Controller's Office and Munitions and Supply about Abasand. Several different observers, including Consolidated Mining and Smelting (CM&S) and the Geological Survey of Canada, wrote that severe problems plagued the facility, its practices and management. CM&S, for one, purchased fuel oil from the plant, but noted that deliveries of finished products were sporadic and haphazard. Meanwhile, other visitors to the site noted several concerns about the operation of the facility. Most pointed out the plant operated on a shoestring budget from the beginning. Renovation suggestions, ranging from the most banal to the reconstruction of the entire plant, received constant support.[20]

Given these circumstances, a fair question might have been: what should come first? Should Abasand design a separation process capable of producing commercial quantities of bitumen from the current operation, or should the company move the facility to a more appropriate location?

The decision placed Howe and the Department of Munitions and Supply in an unenviable position. The development of the oil sands had frustrated the best experts in the field – scientists, miners, and oilmen alike – for the better part of thirty years. Little evidence suggested the federal government could change matters. The Abasand separation process was clearly imperfect at best, incomplete at worse. Could the government afford to invest $10 million of precious capital in a wartime construction project considering that the plant under construction previously displayed little reliability? Against these very troubling problems lay the politically unattractive option of doing nothing. With each passing day, Canada's oil situation grew steadily worse and the cries for the government to improve the situation were becoming more intense. "We need oil at the present time," said Edmonton Member of Parliament Cora T. Casselman in the House of Commons, crystallizing the government's dilemma. "In the Athabaska [sic] tar sands we have a tremendous reservoir of oil ... it seems to me that the federal government might do something toward the development of that resource."[21]

Howe displayed a remarkable degree of restraint in his approach, balancing the need for action against departmental unease over the feasibility of Abasand's operations. Therefore, instead of hastily jumping on or off the oil sands bandwagon, Howe moved with agonizing deliberateness. Confronted with the prospect of choosing between two equally unappealing alternatives, Howe elected to defer the decision altogether. The Minister of Supply and Munitions asked CM&S to decide if the government should continue with the development of a pilot plant.

It was a half-hearted measure that amounted to the government making no policy choice at all. They merely delayed difficult decisions for another day. While this "policy-by-inertia" approach enabled the government to state legitimately that it was exploring the situation, it did nothing to improve Abasand's operations. Privately, Howe hoped the process would prove successful so that he could approach members of the United States government and have them invest in the project. Max Ball quietly worked in the background at the Board of Economic Warfare to try to entice them into the project, but nothing had come of his efforts thus far.

Meanwhile, members of the Opposition grew restless with the government's policy. During question period, they peppered Howe with queries about the progress, or lack of it, on the oil sands. To respond to his critics, Howe made copies of Cottrelle's report available for the Opposition. Perhaps Howe hoped that they could see the same inherent problems with the process known to members of his department. Critically, however, Howe did not share with the Opposition the reports suggesting that Abasand's operations and facilities were seriously flawed. Instead, Howe ended up being too clever by half when he suggested that one of Cottrelle's major criticisms of Abasand – its location – significantly dampened the government's enthusiasm for the project. The reality was that the plant's location was flawed, but only because the plant could not complete the task assigned to it. Even if the entire Abasand facility moved thirty miles to the south, the plant would not suddenly overcome its problems.[22] Thus, the government soft-pedaled its major reservation about Abasand – the inability of its facility to produce an adequate volume of material – to the point where it seemed that the plant's physical *location* was holding up further development.

Although in subsequent debates Howe would constantly try to redirect attention to the flaws identified in Abasand's extraction process, he would never fully succeed. With the debate framed in this fashion, the Opposition and oil sands boosters seized the initiative in the spring of 1942. John R. MacNicol rose in the House of Commons to speak at some length about the tar sands. After surveying the current petroleum situation in Canada, MacNicol turned his attention to Howe's contention that the Abasand facility was too isolated to be an effective wartime project. "It is not far," argued MacNicol, "only 305 miles to Edmonton. A pipeline of 250 miles in length would cost about $2,500,000." Then MacNicol ended his lengthy speech with a powerful argument. "We have in these oil deposits hundreds of billions of barrels of oil, but even if it were only hundreds of millions it would pay this country to develop these resources and become an exporter instead of an importer of oil."[23]

Hoping to sidestep the issue, Howe replied that the government's official policy had changed. Previously, most people regarded the oil sands as a great future reserve. "Today we think of that same area as a source of immediate oil production," said Howe. The minister repeated that, at the government's request, leading engineers from both the public and private sector would look into the situation and report to him. "We have asked all these interests to prepare a report for the government and advise whether we should proceed with a pilot plant or whether the situation is such that we can immediately develop on a considerable scale." [24]

A few days later, another member of the Opposition, Charles E. Johnson, took Howe to task for his perceived inaction. Referring to Howe's prior remarks about an ongoing departmental investigation, Johnson suggested the government was simply dragging its feet. "Surely the work which the Abasand people have done in that area should be sufficient to indicate definitely to the minister whether or not it would be practicable to start a plant," chided Johnson. "It will probably cost a lot of money to establish a plant which can make speedy production of this greatly needed fuel, but the minister would be well advised to continue to keep in mind the possibilities of putting up a plant." Then Johnson returned to Howe's remarks about the plant's location. Edmonton had already provided access to the plant, including paved roads, rail lines, and airports. Surely the presence of this infrastructure eliminated arguments that the area was remote and inaccessible. "The time is long past when this development should have taken place," concluded Johnson. "Even though we are three years late in doing it, I hope the minister will see the urgency and the necessity of starting something immediately." [25]

Perhaps the most damaging performance, however, began later that afternoon when Lethbridge MP John H. Blackmore rose to address the appropriations bill. "I am not by any means satisfied that our resources are being used to the degree to which they could," began Blackmore in one of the most extraordinary speeches ever to take place in the House of Commons about the oil sands. The fate of the British Empire and the United States – no less than six hundred and fifty million souls – depended on Canada's war effort. The departmental committee to which Howe constantly referred was simply wasting time. "What is it about the Athabaska [sic] tar sands that they do not now know?" demanded Blackmore. "It seems to me they could find out very soon if they chose to do so." If the potential cost of developing the oil sands was holding up progress, Blackmore suggested that the minister weigh additional expenses against the prospect of losing the war. "If one Adolf Hitler had control of the Athabaska [sic] tar sands," he asked hypothetically, "does anyone believe for a split second that he would be worrying about the cost of producing the gasoline?"[26]

Clearly, the government had lost the ability to define the terms of the debate and was now on the defensive against repeated Opposition blasts. More important, Howe did not want to reveal everything about the process, given his preference to involve the United States in some fashion. If, on the one hand, he publicly declared that the sands could supply all of Canada's crude oil needs, Canadians would expect the federal government to invest the money. On the other hand, if he seemed too hesitant to begin development, the U.S. Army might be more reluctant to take the project over. Perhaps Howe made a distinction between the potential of the oil sands and the way the Abasand operated its business that was not immediately apparent to Opposition critics.

Howe was willing to concede that the oil sands could produce large volumes of oil some day; he was less certain that they would realize such production with Abasand directing operations. Max Ball was a competent engineer, but he was also away from the facility for extended periods while working as a consultant for the United States government. Without his steadying influence, plant employees fell into sloppy habits. Combined with the jerry-rigged facility, the net result was that production at the facility was uneven at best. Despite signs that the Americans were unhappy with the Canol project, Howe held out hope that the U.S. Army would get involved with the oil sands project.[27] For parliamentary critics, though, Howe's reservations about Abasand amounted to a distinction without a difference. If the oil sands could produce useable oil, then Abasand could produce oil for the war effort.

With public pressure mounting for the government to do something about the dire fuel situation, the Department of Munitions and Supply had to act. In June 1942, the federal government announced that formal steps would begin to develop the oil sands "as a war measure." The first step of the project would be to observe the operation of the Abasand Oils plant. This would "determine whether its design was good enough for the basis of design of a large-scale plant and also to prove up, by drilling and sampling, a body of oil sand sufficiently large to sustain a large-scale development."[28]

The government turned to Consolidated Mining and Smelting Company to evaluate the operation and to report on its findings after the plant closed down in the fall for the season.

In addition, Howe insisted that the Province of Alberta join in the project. On June 23, he wrote to Premier William Aberhart to ask for the province's co-operation with the Abasand effort. "War requirements for petroleum are so urgent," wrote Howe, "that the federal government wish to explore the possibilities for a large scale development in the vicinity of the Abasand operation."[29] After Howe explained the role of Consolidated Mining and Smelting and the overall objectives of the project, he made a special request. Given that the work would be "contingent on the approval and co-operation of your government," Howe proposed a *quid pro quo* arrangement. Since the federal government would proceed, the letter inferred that the province should share some risks. In this respect, the minister requested that the province grant no more leases until Ottawa reached its own decision about large-scale development.[30]

Howe's proposal was troubling to the Alberta government. For one thing, a blanket decision to halt all other exploratory work affected the operations of another company, International Bitumen. Could they "grandfather" International Bitumen into the arrangement, enabling the company to continue their efforts independent of the federal government?[31] Beyond this practical matter lay serious questions about the wisdom of focusing all the research efforts of the country into a single project employing a single separation process. While Abasand had produced some tangible results, they could not guarantee that they had absolutely mastered the oil sands. In contrast, for the Manhattan Project, the U.S. government simultaneously investigated six different processes scientist thought promising.[32] Could the province afford to place all its hopes in the Abasand plant to develop a commercial-scale operation successfully? Perhaps the point was moot because the reality was that the province did not want to be seen interfering with the war effort. Anyway, Aberhart and the rest of his Cabinet understood that the war powers of the federal government precluded the possibility of effectively opposing the federal initiative. If the province chose to withhold its approval, Ottawa could simply invoke its war powers and take control of whatever resources the Minister deemed necessary for the war effort.[33]

Aberhart tried to obtain some face-saving concessions from Ottawa to give the appearance of a consensual agreement between the two governments.[34] On the surface, Ottawa acceded to some of Edmonton's requests, but Aberhart gained little of practical value. International Bitumen could continue its operations, provided it could obtain investment capital from outside sources. Since the province had no budget to fund scientific inquiries, International Bitumen first turned to Ottawa to see if they would be willing to finance a second project. Predictably, the federal government demurred because of its Abasand commitment and International Bitumen struggled to find private investors in Canada or the United States to fund further development. Unable to find a significant source of capital, its operations languished for the duration of the war until revived by the province in 1944–45.[35] Meanwhile, the federal government never formalized the information-sharing agreement. Instead, the personal predilections of Abasand's various project supervisors governed the exchange of information between Ottawa and Edmonton. With little recourse available to it, and with a certain degree of trepidation, the province watched as the Oil Controller's office took control of the Abasand facility in July.[36]

From the beginning, Abasand faced an uphill battle to prove its feasibility. Many problems stemmed from the serious design flaws inherent in the operation. An internal CM&S report concluded that the plant used "second-hand junk" to rebuild a poorly designed facility.[37] Nevertheless, the process of evaluating the facility continued through early 1943. Consolidated Mining and Smelting contacted Dr. Clark at the end of the year to provide an assessment of the operation and to suggest any changes. When Clark visited the plant in August, two things became immediately clear. The first were the overpowering fumes in the building that made breathing difficult and made the whole plant uncomfortably hot. Evidently, the process was losing much chemical diluent and heat. A problem also existed with the pulper at the head of the plant. The intent of the pulping machine's design was simply to separate the oil from the sand, but at Abasand, they also required it to receive and break down large lumps of sand and ironstone nodules produced by blasting. While this would normally have required two different operations, Abasand cut corners by combining the two procedures and did neither properly. Matters did not improve later in the facility. The separation plant, for example, produced more oil than the refinery could process. Thus, from time to time, plant operators shut down the separation plant to allow the refinery to catch up.[38]

At the conclusion of the tour, Clark wasted little time in pointing out the obvious. The facility needed several upgrades, including the complete rebuilding of some sections of the plant. Although CM&S noted Clark's suggestions, they also pointed out that their mandate from the federal government did not include redesigning the facility. Their assigned task was to assess the process currently in place and to decide whether turning the facility into a commercial project was possible.[39]

By far the most important, and challenging, task put before CM&S was to sample and test sand deposits on the Abasand lease. CM&S would have to collect a series of samples from the deposit via a process known as core sampling – where miners would drill into the deposit to learn its composition. What the company did not realize was that they were entering uncharted territory. When Sidney Ells explored the extent of the oil sands in the 1920s, a pick, shovel, and hand-operated augers were enough to collect samples. During the wartime emergency, these methods took far too long to gather sufficient quantities of sand for testing to take place. CM&S began to use power augers to mine the sand. Unfortunately, they soon discovered that they could not drill the oil sands using conventional techniques because of the wear and tear they caused on equipment. The shovel at Abasand could not dig into the sand deposits when the formation temperature was below 45° F because metal drill bits would either wear out or break. CM&S asked Dr. Clark if he knew of a solution to this dilemma. Unfortunately, he did not. The average temperature of the deposit ran at 36° F. While the temperature of a quarry face could rise to 70° F during the summer months when exposed to direct sunlight, the temperature would drop to 45° F at a depth of ten feet from the surface. "This means that at any season of the year at any location," wrote Clark, "power shovel excavation of undisturbed beds is not feasible."[40] Since replacement parts were difficult to find, CM&S decided to design and manufacture its own equipment from scratch, delaying testing even longer. Eventually, by the late summer of 1942, CM&S managed to begin gathering data and drilling core samples.

Consolidated Mining presented its findings to the federal government in the early spring of 1943. Between January 1941 and December 1942, a series of independent economic assessments provided by the Geological Survey, the Department of Mines, and CM&S, respectively, estimated that production costs at Abasand were steadily dropping. They ranged from a high of $2.00 per barrel to a low of $0.86 per barrel while world prices hovered around $1.60 per barrel. Despite these encouraging signals, CM&S acknowledged that wartime demand and strategic need would still determine whether the government would continue to invest in Abasand.[41]

When the report turned to an overall assessment of Abasand, matters were less glowing. Problems turned up in nearly every phase of the Abasand process. The separation plant worked for only approximately seventy days in 1942 and produced just less than 10,000 barrels of crude. 10,000 barrels was the same total amount *per day* the federal government figured would be necessary for a plant to be of value for the war effort. The refinery worked for eighty-two days and produced 712 barrels of oil. However, as CM&S noted, Abasand itself used 43 per cent of the plant's overall production to run the plant, making Abasand its own best customer. The report also noted the need for large amounts of capital to fund additional research and to finance a complete overhaul of the facility's physical plant – estimated at $268,000. In total, the report proposed that Abasand required an additional investment of approximately $1 million before the plant would produce fuel oil in any sufficient quantity.[42]

REORGANIZING ABASAND

Meanwhile, Canada's fuel situation deteriorated even further, making Howe's decision critical. Ball and his group met with the federal government and admitted that they could not cover the costs of the upgrades by themselves. With little alternative available, Ball agreed to turn the plant over to Ottawa for the duration of the war.[43] The terms of the takeover were straightforward. The reorganized Abasand Oils Limited installed a board of directors approved by the federal government. If Ottawa abandoned the project, ownership of the property and facility would revert to the original firm. However, Ottawa would not compensate Abasand Oil for the appropriation of its plant or any of its equipment. The exchange seemed fair since Canadian taxpayers would foot the bill for reorganizing the company, upgrading the facility, and bringing a pilot plant on stream.[44]

Despite some grumbling by members of the Opposition, little question remained that the government would redesign and reconstruct the entire Abasand operation.[45] The only question that remained would be to find an appropriate company to run the facility. The only expert on the oil sands working in the federal government, Sidney Ells, had no experience with commercial operations. No one else in the Department of Mines displayed very much interest. In order for the government-run plant to succeed, many believed it would now need to draw on the expertise of the private sector.[46]

The most obvious candidate to take over the plant was Consolidated Mining and Smelting. Despite valuable and ongoing interest, research, and suggestions from CM&S, it does not appear that the government seriously entertained the company's proposal to take over the plant. "The prejudice against the miners also would appear to have included either fear of or disdain for the impressive financial strength a subsidiary of the C[anadian]P[acific]R[ailroad] could bring to the sands," concluded one historian.[47] Instead, Howe reorganized Abasand's board of directors and included representatives from Nesbitt Thompson, and Canadian Oils, two of the original backers of the Abasand project. If the federal government was going to back the project financially, it wanted to have input into the company's operations. Thus, they appointed representatives from the Department of Munitions and Supplies Oil Controller's Office to the board to oversee developments. "We had Ball and associates undertake the work in 1935," noted Howe in the House of Commons. Although they worked very hard, there was nothing more tangible than data to show for their efforts at the end of 1942.[48] Clearly, changes were in order.

However, while federal officials wanted to oversee the project, they realized that they did not have the expertise or technical knowledge to run the plant themselves. Instead, they invited Earl Smith from Canadian Oils to serve as superintendent of the facility.[49] The government immediately pledged $500,000 to rehabilitate the plant in a joint venture between the federal government and the private sector. "What we are trying to do is to find a process for developing the sands," said Howe. "I sincerely hope we shall be able to find the solution of many problems by the autumn of this year ... but one would have to be very optimistic to believe that we shall be able to start any large-scale construction before the spring of 1944."[50]

Meanwhile, Opposition members in the House of Commons were not at all pleased with the thrust of the federal effort. Some members began rehashing old fears of a sinister international cabal seeking to control and dominate the oil industry to the detriment of consumers. "We want to get out of the hands of the big fellows," said John MacNicol. "We want oil produced in this country and we want it produced at the right price ... I would feel more confident if the oil companies did not have a thing to do with the development of oil from these tar sands. I would much rather see this carried on by people who are not personally interested in oil." Howe did his best to reassure MacNicol that Canadian Oils Limited was not sinister. "Our purpose in sending in a new management, or a partly new management – because the old directors are still associated with the company – was to give a more practical turn to the operations in that field. The reason that the Canadian Oil Company was chosen is that it is a purely Canadian company and the one oil company I know of that believes in the future of the tar sands," replied Howe.[51]

Despite Howe's assurances that more experienced Abasand employees would remain working on the project, the reconstituted Abasand clearly bore little resemblance to its predecessor. Soon after the government acquired the Abasand Plant, Ball returned to the United States to work for his own government and severed nearly all his ties with the Abasand plant. More alarming, wrote Clark later, was that "most everybody connected with the development of the company disappeared." Some have since speculated that the company's American roots inspired a housecleaning at Abasand as Howe tried to "Canadianize" the project. However, after considering Howe's longstanding attempts to enlist the support of the United States government in the oil sands project, this seems highly unlikely. In fact, Howe actively followed and encouraged Ball's early overtures to the United States government.[52] A more plausible explanation relates to the mismanagement of the plant from its very inception. The number and frequency of mistakes, blunders, and sloppy habits known to the Department of Munitions and Supply make this explanation more likely. If Ball and his crew could haphazardly spend investment capital from private backers, the Department of Munitions and Supply must have feared what would happen to government funds.

While such an explanation is easy to understand, and perfectly justifiable under the circumstances, it ignores one essential factor: experienced personnel were an absolute necessity for the plant to operate smoothly and efficiently – even one as poorly constructed as Abasand. Repeatedly, the oil sands displayed properties baffling to miners and oilmen alike that only years of work and experience in the field could conquer. In a series of puzzling and questionable moves, the Abasand operation under the direction of Earl Smith steadily eliminated the reservoir of available knowledge.

Since its inception, the Research Council of Alberta had maintained cordial working relationships with all of the private companies operating in the oil sands. While Ball and the Alberta Research Council had a stiff, almost perfunctory, association with one another, they would still share bits and pieces of information. Once the federal government completely took over the Abasand facility in May of 1943, however, all contact between Abasand and the Research Council of Alberta abruptly ended.[53] Clark still received independent inquiries from officials associated with the plant, but the scientist was unable to see the revamped facility. "I do not know much about what Abasand is doing as relations are not particularly cordial," wrote Clark to E.O. Lilge of CM&S in July.[54] Whatever information he managed to glean came through private letters and conversations rather than witnessing plant operations firsthand. For the first time, the Province of Alberta found itself on the outside looking in on a high-stakes game over the province's future. "The prospects of a bituminous sand boom depends largely on the answer that the government effort turns up," acknowledged Clark with some trepidation.[55]

Perhaps more unusual was Smith's decision to exclude the Department of Mines own expert, Sidney Ells, from deliberations. Ells could only gather information from his network of friends in Fort McMurray about Abasand. Nevertheless, in public speeches Howe maintained that Ells was heavily involved in the government's effort. Opposition members sharply rebutted Howe's contention and argued that, although Ells was nominally involved with the project, Abasand ignored his advice. "I was there," said Mr. MacNicol, "and I know how little Doctor Ells had to do." Therefore, at the most crucial point of Abasand's operations, the federal government neglected the best experts in the field.[56]

Other personnel turmoil ravaged the plant. Several labourers and engineers left Abasand to find greener pastures working for the United States Army on the Canol Project. Employees reported that the Americans paid more money, offered better working conditions, and even offered to drive workers to and from work. Beyond these issues, the Abasand facility did not enjoy the privileged status of other wartime endeavours and had to scrounge for labour, equipment and supplies. Faced with such stiff competition that the company could not match, Abasand's workforce suffered accordingly. Despite new facility leadership and direction, many of Abasand's old problems continued unabated.

Smith did not plan to make any major changes to the facility despite repeated internal warnings that the facility's flaws were "possibly mortal." Rather than improve the separation and refining functions of the facility, Smith devoted most of his efforts to improving buildings not essential to the production of bitumen from the oil sands. Meanwhile, Smith displayed little ability to get the most out of Abasand's employees and gradually began alienating most of the facility's key people. J.M. McClave, part of Ball's original group and designer of the plant's separation facility, resigned. Although McClave still occasionally visited the site, Abasand lost its leading expert on the design and function of its separator. Reports began to circulate in August that there were instances of deliberate sabotage at the plant. They alleged that an unidentified worker placed sand in the lubricating oil used in some vehicle engines. In mid-September, word of the unrest and lack of progress found an audience in the Oil Controller's office. George Cottrell noted that even Smith's superiors at Canadian Oils acknowledged that work on the Abasand Plant lagged far behind. Cottrell suggested a change at the highest level of the Abasand management team. Anticipating the struggle, Smith stepped aside and Boyd Webster of the Oil Controller's office replaced him on September 30.[57]

While Smith was an oil man, Webster was an engineer. The difference between the two became apparent immediately. Well aware of the problems that had plagued the facility since the outset, Webster decided to rectify the situation. Webster addressed deficiencies in the facility's efficiency by overhauling the entire management structure of Abasand and revising the work schedule. Unfortunately, Webster also confronted the fact that the renovations undertaken in May were incomplete when he took the plant over in September. By the time he carried out his changes and completed the plant upgrade, Abasand had lost the 1943 production season. In addition, more money would be necessary to upgrade the plant. Early estimates calculated that the federal government would have to invest at least another $778,000.[58] By the end of 1943, the Abasand facility teetered on the verge of disaster. Officials in both the Department of Mines and Resources and the reorganized Abasand Oils believed outside expertise would be necessary to improve operations. They approached General Engineering to redesign the separation plant and Imperial Oil to retool the refinery.

On the surface, General Engineering's selection seemed a good choice. The company's experience lay with designing surface plants for nonferrous mines all across the globe. However, the Athabasca sands were an entirely different creature and one with which the company had no experience. Thus, the new managers tried to re-establish Abasand's links with leading researchers in the field, particularly those associated with the Alberta Research Council. In early December, engineers from General Engineering met with Dr. Clark to discuss the oil sands and the hot-water separation process. What General Engineering revealed to Clark was not encouraging. "Mr. Hamilton admitted, frankly that he and his company knew nothing about tar sand separation plants," wrote Clark in an *aide-memoir* after the meeting. "From the engineering standpoint, what had been planned by the old Abasand and revised under the Earl Smith's regime was bad engineering." The new management scrapped everything.[59]

But Hamilton admitted that General Engineering also faced intense pressure from the federal government to use the same basic process and equipment previously employed. The Department of Mines insisted that the Abasand process worked and taking the time to find another process would waste valuable time and money. Thus instructed, General Engineering's own experts studied the flow sheets and plans from the previous regime and concluded that the equipment available was unsuited for the task. For Clark, this revelation confirmed his darkest fears about happenings at the plant. The federal government's stubborn insistence on using Abasand's design was the latest manifestation of a "lack of curiosity" plaguing Abasand's operations from the beginning. "We knew that the Abasand scheme did not make use of what was known about bituminous sand separation," wrote Clark. The present federal effort, he concluded, seemed like nothing more than a risky throw of the dice. Clark and other members of the Research Council reached the pessimistic conclusion that the gamble would fail and jeopardize the future of oil sands development.[60]

"SABOTAGE!"

Whatever goodwill existed between the province and the federal government evaporated like mist before the sun. Allegations of mismanagement, heretofore whispered about in Fort McMurray and Edmonton, soon found full voice. Provincial politicians began to investigate the happenings at Abasand and did not like what they found. The Alberta Minister of Public Works, Mr. W.A. Fallow, led the charge, publicly alleging sabotage at the facility in a November speech. The national press assailed Fallow's remarks as patently false. However, enough information about events at the plant, particularly the deliberate sanding of lubricating oil, circulated to cause the federal government some unease about future operations.[61]

Finally, on January 7, 1944, Premier Manning wrote a concerned letter to C.D. Howe about the reports emanating from Fort McMurray about the mismanagement of the Abasand facility. Manning could not escape the conclusion that Abasand's progress fell far short of expectations.[62] Particularly galling to Manning was the fact that the province set aside a large area of the oil sands region with nothing to show for its troubles.

For the Province of Alberta, the oil sands were not just the answer to a wartime shortage of fuel. They represented the potential of a multimillion-dollar energy industry in a predominantly agrarian prairie province. Ottawa barred other entrepreneurs willing to explore and develop the sands from initiating operations.[63] If the federal government could not develop the resource effectively, the province would turn the lands over to someone else.

Still, to be sure, Manning could only push Howe so far on these issues, particularly in wartime with scarce resources for research available. "We are determined not to interfere with any activity which the federal government feels is in any way essential to the maintenance and expansion of Canada's war effort." Clearly, however, the province could not wait much longer. Given the federal government's perceived lack of progress or interest in the project, Manning concluded his letter to Howe with a warning that the province might initiate its own project. "We do not feel justified in refusing [any] longer to grant crown leases for the development of tar sand deposits outside that area," wrote Manning. The provincial government therefore asked Ottawa to release it from the understanding of July 3, 1942.[64]

In his response, Howe wasted little time in addressing the premier's concerns point by point. Howe first took issue with Manning's assertion that the Abasand plant had not yet produced any tangible results. "The fact that our work is not completed arises from the disappointing results so far obtained," claimed Howe.[65] The federal minister contended that a "lack of uniformity" in the oil sand deposits introduced unforeseen problems with the separation process. Then Howe acknowledged that the separation process tested at the Abasand facility was highly inadequate, something already known to the Research Council's top scientists but ignored by federal authorities. Howe admitted "our work with the existing operating unit disclosed that the methods being employed are not satisfactory for large scale commercial purposes." Only now was the redesigned operating unit capable of turning out suitable samples. The conclusion was inescapable. Abasand would fail to deliver usable oil for the current wartime emergency.[66]

More alarming was Howe's next contention. The minister suggested that the commercial exploitation of the oil sands was not possible due to the composition and properties of the oil sands rather than a faulty separation process. "The reports that I have received to date do not indicate that we can be optimistic about obtaining a large commercial outlet for this product," concluded Howe matter-of-factly. Then, Howe revealed that he doubted they could ever develop a separation process to make the oil sands a viable energy source. "I hope that we can carry on our investigation to a point that will settle *for all time* the possibilities of the tar sands as a source of petroleum products," concluded Howe. Ottawa would obtain such information by the fall.[67] Meanwhile, Howe assured Manning that Ottawa would release the province from its obligations under the July 3 agreement. But Howe then threatened to withdraw all federally funded research if new lessees used the same processes investigated at Abasand before the government could announce its findings.[68] The implication was clear. Either the province could allow the federal government to continue its inquiries unimpeded or it could watch the oil sands remain idle.

In all fairness to Ottawa, federal officials were well aware that the plant had problems in the previous year. They were taking steps to rectify the situation and there seemed little point in rehashing old grievances. New managers from General Engineering were in charge of the facility. They hoped to have the new separation plant up and running again by April. However, politicians in Edmonton cast a decidedly darker and more cynical eye over the situation. If the oil sands development failed to provide usable fuel, it was because of incompetence or deliberate sabotage by unnamed parties who did not want it to succeed.

Perhaps what officials in Ottawa failed to recognize was that Alberta's government simply could not allow development of the oil sands to fail so spectacularly on such a large public stage. Research into the oil sands was now well into its third decade, but federal authorities stubbornly insisted on employing separation techniques doomed to fail from the outset. Instead of tapping into the large body of experts on the oil sands Alberta possessed, Ottawa seemed content to bring in outsiders. Although they often came with impressive credentials, Ottawa's experts were, nonetheless, neophytes when it came to dealing with the oil sands. The Second World War provided a window of opportunity to establish the viability of the oil sands. With the federal government in charge, it proved a giant flop.

Howe's veiled threat to withdraw all funding stirred members of Alberta's Legislative Assembly. Several ministers assembled all available information on the federal government's activities at Abasand. The Minister of Public Works, W.A. Fallow, remained in close contact with Karl Clark through the Abasand years. The previous September, amid the allegations of sabotage and the dismissal of Earl, Fallow made it a point to seek Clark's opinion about the province's next move. Now Clark gave Fallow powerful ammunition to launch an all-out attack on Abasand in a scathing indictment of Abasand's operations since 1942.

In a letter addressed to Fallow, Clark outlined the torturous relationship between Abasand and the Research Council of Alberta, including the abrupt end of relations in 1943. "The main usefulness of this account could be to help you relate what you plan to say with general background of fact and to avoid giving openings for a comeback," wrote Clark.[69] Next, Clark provided a series of newspaper references about North America's petroleum shortage and the potential of the oil sands as a supplementary source of fuel. This, argued Clark, would establish the public stage the oil sands found during the wartime emergency. Finally, Clark furnished the minister with a list of specialists and experts dismissed or who resigned since the federal takeover. "It looks very much as though it was the studied policy of this company as now organized to have nobody about who has had any experience with bituminous sands," offered Clark.[70]

Clark knew that the information he provided to Fallow was powerful and highly inflammatory. However, the comments and suggestions contained in his report did not stray far from what Dr. Clark had been saying and writing about Abasand for quite some time. On this point, Clark wanted to be clear. Fundamental problems existed with the Abasand facility. His job, however, was not to point a finger and assign blame as much as it was to help the plant succeed. Thus, he told Fallow "it would be advisable, I believe, to keep my name out of an attack on Abasand. If the company now wishes our co-operation … I am the one that will have to be on personal terms with the company and its men."[71]

The public at large widely knew of provincial displeasure with the Abasand project. Specific grievances, including Fallow's intention to demand a Royal Commission investigation of Abasand, circulated freely in local newspapers. On March 13, when Fallow rose in the legislature and charged that "an apparent waste of public funds" occurred at the Abasand plant, it was little more than what people expected. Then, in a shot at the federal government, Fallow asserted that the gross mismanagement of the facility led to public fears that Ottawa deliberately and systematically sabotaged the development of the oil sands.[72]

In his remarks, Fallow touched on several subjects before turning to more recent events. He recounted the dismissal of CM&S from operations, the rise and fall of Earl Smith, and the eventual appointment of General Engineering to clean up the mess. Although Fallow never explicitly or directly accused Ottawa of sabotaging the plant, the implication was clear for even the most casual observer to see. "In summing up, Mr. Speaker, may I say that the pathetic flounderings of this gang of pussy-footing amateurs beggars description. In all my experience, I have never seen a more ridiculous attempt to pull the wool over the eyes of the people." [73]

What outraged Fallow even more than the gross mismanagement of the facility was the stigma that this half-hearted effort might forever attach to the oil sands. Had the federal government given the project an honest effort and come up wanting, perhaps the province's reaction might have been different. Yet what was so galling was an unyielding belief that the federal government had not given the oil sands a fair chance to prove its value. In the midst of a national emergency, rank amateurs appeared to manage the federal oil sands initiative.

With no shortage of suspects, perhaps it was inevitable that the public fallout tapped into Alberta's longstanding resentment, suspicion, and mistrust of "big oil," generally, and the Eastern establishment specifically. The involvement of the federal government and Alberta's longstanding bitterness toward central Canada served as the prism through which people interpreted news coming out of the Abasand facility. Thus framed, the debate assumed a bitterness and rancour totally divorced from actual events. It no longer mattered whether or not the Abasand facility could separate oil from the sand. In the public mind, the only important fact was the federal government's involvement.

When the Opposition raised the issue in the House of Commons a week later, Mackenzie King's government was on the defensive. After nearly two years of serving as the public mouthpiece for the Abasand effort, the Department of Munitions and Supply belatedly divorced itself from the project. "I have nothing to do with [Abasand]," said C.D. Howe before adding for the first time that the oil sands effort was an initiative of the Department of Mines and Resources. [74] Undeterred by Howe's comments,

John MacNicol continued to read the text of Fallow's remarks into the official record much to Howe's obvious displeasure.

Howe tried to prevent Fallow's charges from becoming part of the public record. He claimed Fallow's accusations undermined the integrity of the House of Common and implored the Speaker to prevent MacNicol from continuing his speech.[75] "I want to tell the minister he is altogether too combatant," chastised MacNicol. "He is too fond of arguing with anyone who says anything about his department."[76] If Howe were not directly involved with the plant, asked MacNicol, "why then does he want to rise and interrupt my remarks?"[77]

Realizing that he had painted himself into a corner, Howe resumed his seat. Howe only interjected from time to time that Abasand did not come under his department's direct control. Passions on both sides of the aisle ran high on this debate as Opposition members tried to establish government involvement with the project. Meanwhile Howe flatly denied any direct government involvement with the plant and its policies. The exchange finally ended when Howe asked the Speaker for more time to "examine the statements and obtain the opinions of my experts. I shall try to make a statement on this matter before the debate ends." Finally satisfied after nearly an hour of debate, the Opposition allowed the committee to take up new business.[78]

Howe never did make his promised statement. In the intervening weeks, authority for the Abasand project mysteriously shifted department portfolios from C.D. Howe to Thomas Crerar in Mines and Resources. Although Howe still made the important decisions regarding Abasand, he publicly distanced himself from events.[79] Crerar enjoyed a long and diverse political career, first as leader of the Progressive Party in the 1910s and then as a Liberal after 1929. His sudden emergence as the government's public representative on the oil sands altered the parameters of debate. Where Howe was inclined to behave like a combative businessman when questioned, Crerar had more of the politician's polish about him. He understood the nuances of parliament and was less likely than Howe to construe probing questions as a personal attack. More important, Crerar deflected attention from Howe and enabled the government to distance itself from earlier decisions.

One such example occurred on May 23, 1944, when Thomas Crerar rose in the House of Commons to request an additional $500,000 appropriation for Abasand. Crerar gave a passing acknowledgment to the fact that the Department of Munitions and Supply initiated the project. However, the bulk of his address attempted to minimize the government's involvement in Abasand. The most he was willing to concede was that Abasand was a joint venture between government and the private sector.[80] Crerar maintained that every alteration to the plant took place only after consultation between Abasand's technicians and the Department of Mines and Resources.[81]

While the Opposition was prepared to grant Crerar a certain degree of latitude on some issues, it did not accept the government's version of events. "The Abasand company with which the government is dealing is not the Abasand company which built the 1942 plant," pointed out John MacNicol. "The government took over the works from the Abasand company, but there is no official belonging to the original Abasand company working there today that I know of." MacNicol contended Ottawa, not Abasand, operated the plant.[82]

Because of the significant amount of public money spent on the project, the Opposition added their voices to those demanding a formal investigation. The simmering debate, which already possessed an undercurrent of hostility, soon boiled over. Backbenchers and Opposition MP's traded barbs, cast aspersions, and hurled insults across the aisles – all properly cloaked in the mock politeness of parliamentary rules.[83] Even a brief recess did nothing to mitigate the sour and petulant atmosphere in the committee room. The government was clearly on the defensive and began to lash out. Alberta's Minister of Public Works, W.A. Fallow, got his information from "drunken men"; an Opposition MP who questioned the government's actions "was elected in his own riding in 1935 through a fluke" and was now seeking publicity to help with re-election; Max Ball, who designed the original Abasand plant was not an engineer at all; the Province of Alberta had jurisdiction over Abasand and should have conducted its own inquiry.[84]

The litany of accusations and counter-accusations mounted until finally Charles Johnston stated the obvious: "The minister must accept responsibility for the things done in his department."[85] Again, Crerar took the offensive, parsed words, disputed figures, and generally sought to reduce the government's exposure. While closing debate on the issue, Crerar ended the session with a monologue that left little doubt how the government saw the developments related to Abasand.

> The Alberta government could have gone ahead fourteen years ago and aided in the development of these sands. It could have put in its own plant. It could have used public money and bonused the industry, encouraging the industry to go on and develop the sands. But it did not do so; and now, when this government is making an honest effort to find out the commercial possibilities of these sands, all we get by way of thanks in return is the sort of speech that Mr. Fallow made in the legislature and the kind of speech my honourable friend made here this evening.[86]

Crerar's self-serving version of events, including the abbreviated history of the oil sands, did not attempt to refute the main thrust of Alberta's complaints. Nevertheless, suggestions of a vast conspiracy to prevent development of the oil sands by a coalition of business and government helped discredit those complaints. Both sides soon wearied of the fight and the matter of appointing an independent inquiry quietly disappeared into the shadows. The Province seemed content to let the federal government play out the Abasand melodrama. In the interim, provincial politicians focused on their own re-entry into the oil sands and allocated $250,000 to construct a new test facility.[87]

Abasand did not disappear quickly. General Engineering continued its research through 1944 and 1945 and pursued alternative separation methods. While the plant produced some tangible results, another fire in June 1945 razed the plant again. When the federal government tabled its final report on Abasand in December of 1945, little in it surprised observers. While Ottawa agreed that the separation of bitumen from the sands was profitable, it also concluded the federal government should not maintain its presence in Northern Alberta. The Province of Alberta's commitment to the development of a new test plant made the federal government reluctant to compete with the province.[88]

Other reasons prompted the federal government to abandon the oil sands project. Before the end of 1945, Sidney Ells announced his retirement from the Department of Mines, effectively removing the federal government's only oil sands expert from consideration. Without Ells prodding his superiors, the federal government displayed little interest in actively pursuing oil sands development for the next thirty years.[89]

Overall, C.D. Howe's accomplishments as Minister of Munitions and Supply during the Second World War should not be diminished because of federal mismanagement of the Abasand facility. Abasand clearly had more than its share of difficulties well before the federal government became involved with the plant. Perhaps the passage of time permits more sober assessments of Abasand's abilities than those that appeared in contemporary accounts. Clearly, Abasand was neither the panacea its supporters claimed, nor the pariah it became under federal control. While the plant enjoyed modest success, the oil sands contained enough mysteries to confound miners and refiners alike. The development of an adequate separation process still eluded researchers and prevented large-scale commercial development. Until a plant and separation process proved otherwise, the potential of Alberta's oil sands would be more important than its actual production.

Ultimately, however, what distinguished Abasand from other scientific projects during the war is that, in the final analysis, this was not a war-winning weapon. If Canada did not succeed in developing the oil sands, it could continue to rely on conventional sources. Canada turned to the United States to meet its fuel demands. The same was true for the British who relied on America and to a lesser extent the Middle East, for oil imports. Perhaps the only major combatant who turned to synthetic fuels was Germany. The Nazis recognized from a very early date that their war effort was entirely dependent on the success or failure of I.G. Farben's synthetic fuel program. Toward the end of the war, when Allied soldiers encountered German units supplied entirely with synthetic fuels, scientific and technical experts quickly organized to gather as much information on the German synthetics program as they could. It is clear, however, that given the money, research, and investment necessary to construct a commercial plant, synfuels were not alternatives for the Allied war effort. While the British tinkered briefly with adopting a synthetic fuel program to convert coal to oil, they ultimately rejected the solution as impractical. Not only would a synthetic fuel program be capital-intensive, but the British also assumed that relying on imported conventional oil would better serve national security concerns. Importing large quantities of conventional crude from a multitude of ships, operating out of a number of ports, would be much safer. Furthermore, it would prevent British supplies from being concentrated in a few large, easily identifiable – and easily bombed – hydrogenation plants. Although several different processes showed varying degrees of promise, oil companies placed greater emphasis on obtaining and developing conventional crude sources.[90]

Canada's synthetic facilities, particularly those in Northern Alberta, were well beyond the reach of German and Japanese bombers. Why, then, did the development of the oil sands fail so spectacularly during the war? Undoubtedly, the simmering conflict between oil sands investigators in the federal government and the Province of Alberta played an important part. Although Clark and the tandem of Ball and Ells displayed the utmost of professional respect for each other, the competition between them clearly weakened their overall position. Instead of combining efforts, the simmering conflict between the two groups allowed the oil sands issue to become subsumed within the federal-provincial battleground over natural resource development.

CHAPTER 3

Picking Up the Pieces: Reclaiming a Provincial Resource

The discovery of the Leduc field will lessen the interest in tar sand development, I have little doubt. But I do not think that it lessens the importance of tar sand development. I do not think there is any use trying to make out that the tar sands are other than a "second line of defence" against dwindling oil supplies. It is important that a feasible plant for dealing with the tar sands be demonstrated and that the cost of the tar sand oil be established. Otherwise we do not know whether tar sands are a second line of defence or any defence at all. – *Karl A. Clark, July 8, 1947*[1]

The onset of the Cold War at the end of the Second World War was a mixed blessing for synthetic fuels producers. On the one hand, the era of permanent crisis lent an air of importance to the production of fuel from unconventional sources. On the other hand, the Cold War brought global responsibilities and obligations for politicians and bureaucrats in Washington to consider. The Second World War displayed unequivocally the importance of fossil fuels for modern combat. In an increasingly mechanized world, military analysts feared that an enemy would target sources of imported petroleum to bring American forces to a grinding halt. As America weighed its new responsibilities against its available supplies, the administration of President Harry Truman evaluated the country's energy resources. Soon, the administration began discussions about setting up a full-scale petroleum strategy. As part of this effort, the development of commercial synfuel projects therefore became a priority for the Truman administration. Meanwhile, as unquestioned hegemon of the western alliance, America now had the responsibility of ensuring that its allies had adequate supplies of raw materials as well.

Given the rapid market expansion for oil in Europe's economy, questions about the security of Middle Eastern crude supplies took on added importance. Strategic planners in both the United States and Great Britain believed that access to the oil of the Middle East was essential to the development of the west's security. Officials in the State Department began to consider the importance of oil to the industrialized economies of the west. They correctly assumed that America's leadership role would depend, in large part, on the ability of the United States to provide access to vital raw materials. Furthermore, by providing access to markets, the Americans would also ensure the gratitude of producing countries. Even before the war's end, State Department planners began evaluating America's future role. "Is our foreign oil policy aimed at maintaining a strong American influence in the distribution of oil throughout the world," asked one State Department official, "thereby both serving our own national interests and keeping us in position to serve weaker nations according to our ideals – or to allow our present and historical strength in this field to pass into other hands?" [2] Gradually, from 1943 forward, the approach taken by the State Department emphasized serving the needs of both producers and consumers.

Meanwhile, Canadian politicians and businessmen struggled to come to grips with the fallout from the Abasand debacle and Canada's role in the postwar world. The federal government believed that its wartime experiment with the oil sands was an unmitigated disaster and that the future of Canada's synthetic industry was, at best, clouded. Provincial politicians, however, did not share Ottawa's pessimism. Convinced that Abasand failed because of bureaucratic incompetence, the province was determined to prove the feasibility of its oil sands as an alternate fuel source. Before the debate closed on the failed federal effort at Abasand in 1944, the province pledged itself to the construction of a demonstration plant. Renewed provincial efforts would, hopefully, attract further private investors to the field. "If this resolution is carried out," said one cabinet member, "we are going to make an honest experiment to find out what the oil sands hold for the people of Alberta. If what we find out is favourable, let us go the whole way."[3]

The political and economic dynamic of the oil sands industry was on the cusp of a radical change. The driving force behind the oil sands from 1920 to 1945 was the simple fact that Canada lacked a large domestic petroleum industry. In 1946, Canadian wells still only produced enough oil to cover 12 per cent of Canada's annual consumption. Imports of foreign oil supplied the remaining 88 per cent. As oil companies continued to search for producing fields without much success, the oil sands represented Canada's best chance at energy self-sufficiency. Thus, the Alberta government continued to invest in the oil sands not only because it believed in the project's feasibility, but also because there was little alternative. As of 1945, the sands represented the province's best chance to diversify its economy and to one day produce enough energy to supply the country.

However, what would happen to oil sands development if a conventional producer discovered a major oil field within the province's borders? Development of the oil sands did not preclude the search for conventional oil and gas. In 1945 alone, the Alberta government revealed that the search for fossil fuels resulted in more than 3,500 petroleum and natural gas leases for the conventional industry.[4] Improvements in exploration techniques, drilling, geophysics, and geology increased the likelihood of discovery and shrank the oil sands' margin for error.

Uncomfortable questions remained for advocates of the alternative fuel source. Would oil sands investments continue if conventional producers found a major field? More important, would the oil industry continue to pursue development if there were significant conventional reserves to exploit instead? Clark, for one, anxiously watched developments in the provinces and continued to press the province to assume the role of industry catalyst. The scientist maintained that the Alberta government must now show that development would be profitable, since the mechanics of the separation process worked.[5]

CANADIAN OIL IN THE POSTWAR WORLD

Of more immediate concern to Ottawa was figuring out where Canada fit into the postwar constellation of powers and determining what Canada could do to support its allies. As a "middle power," the country seemingly had little to offer in the realm of great power politics that now defined the postwar era. Lacking large conventional military forces and nuclear capabilities, the country exercised little influence in constructing the postwar world order. "Canada was no more than a vocal bystander in great power debates," noted Thompson and Randall. "As such, Canada did not participate at Quebec, Moscow, or Dumbarton Oaks, except as a diplomatic whisper in the British ear."[6] Even if Canada did not rank as an important consideration to American policymakers, Canadians regarded their relationship with the United States as a priority. The Second World War ushered in a new era in Canadian-American bilateral relations. Increased contact between the two nations inevitably influenced the growth and development of the Canadian oil industry. Furthermore, the emerging Canadian-U.S. relationship would have an important impact on the development of the Canadian oil sands.

Ottawa's attitude toward the Canadian oil industry generally, and the oil sands between 1945 and the early 1970s in particular, was one of benign neglect. Given the low levels of production coming from Alberta's oil fields, there seemed little reason for federal politicians to press for the development of a national energy strategy. Since access to cheap, imported fuel was important to Eastern Canadian markets, the development of Alberta's oil patch was of little consequence on the national political agenda. Production and shipping costs for Alberta's crude prohibited any serious attempts to implement a national oil strategy. Ottawa displayed little enthusiasm for telling Eastern Canadian consumers and refineries that they would have to pay more than the world price to sustain Alberta's oil industry.

Nevertheless, not all the developments in Alberta's oil patch were negative. The potential battle between the province and the federal government over control of the oil sands failed to materialize. Ottawa's involvement with Abasand soured the federal government on pursuing any further synthetic fuel developments and the province assumed the dominant position in oil sands development. This did not mean that federal concern about natural resources entirely abated. Nevertheless, Edmonton, not Ottawa, clearly held the hammer in determining the pace and scale of Canada's synthetic fuel industry.

If political control over Canada's oil industry came under provincial jurisdiction, effective day-to-day control centred in corporate head offices of several American-owned multinationals and their Canadian subsidiaries. Despite later writers who believed that American control of the Canadian oil industry was the product of some vast conspiracy, the Canadian industry evolved into a "continental" system. The distinct lack of marketing alternatives made the continental system inevitable. Very few investment dollars from Eastern Canada made their way out west. Instead of direct investment in the oil patch, federal government policies created an investment climate where American multinationals were encouraged to expand their operations in Western Canada. Large American multinationals, like Standard Oil of New Jersey, Chevron, and Amoco, with their large transportation and refining capacities enabled Alberta's growth to take place.

Furthermore, developments in the Canadian oil patch reflected the experience of the American oil industry. Particularly successful were those companies with familiarity of the oil fields of West Texas and Oklahoma. The geology of those fields bore a striking resemblance to the Western Canadian sedimentary basin. In fact, the basin is a northwesterly extension of the Interior Plains of the United States with similar rock formations and geological history. As Eric Sievwright explains, the only significant difference between the two areas is that a thick layer of glacial till covers most of the surface of the Western Canadian basin. "This till and the eroding action of the glaciers that produced it have eliminated almost all surface evidence of structures or folds in the underlying sedimentaries which might provide traps for migrating oil and gas."[7] It only stood to reason that American companies who had enjoyed success at developing similar fields south of the border sought to develop the fields in Alberta.

Not surprisingly, all these factors together influenced the growth and direction of the Canadian oil industry. The significant presence of American-owned multinationals ensured that Alberta's industry would supply markets in the United States. The presence of the large and high-priced U.S. market, serviced by short pipelines, meant that the Canadian oil industry developed along "continental" lines (north-south) rather than along "national," or east-west, lines. Whether they intended to or not, Canadian politicians passed the initiative for development to both the province and American multinationals. "Canadian policy makers were simply unwilling to make a commitment and to pay the costs necessary to achieve greater national independence in the oil issue area" summarized Kohler.[8]

In part, these developments also reflected the larger economic patterns of the postwar years. Some Canadian historians have suggested that the country's inexorable push toward the United States owed a great deal to Mackenzie King's friendly relations with Franklin D. Roosevelt. Others have convincingly argued that the closer relationship owed more to postwar realities than to personal preferences. Before the war, Canada gradually moved closer into economic orbit of the United States, and the emerging bipolar Cold War simply hastened the change. Canadians greeted the immediate end of the Second World War with an extended shopping spree, as did consumers in the United States. Consumer spending rose 11 per cent in 1946 and an additional 7 per cent in 1947.[9] Sales of automobiles, for example, sharply increased. In 1941, only 36.7 per cent of Canadian households owned a vehicle. That number increased to 43 per cent in 1951 and leaped to 68.4 per cent in 1961.[10] "Political stability and the perpetuation of the Liberals in power depended on the Canadian economy operating at a high and stable level," wrote Robert Bothwell. "That level, economists and politicians agreed, hinged on trade."[11]

Additional spending increased American imports into Canada to $1.4 billion, pushing Canada's trade deficit with the United States to $500 million in 1946. The trade deficit with the United States did not worry the government as much as the British decision to continue the freeze on convertibility of sterling into American dollars. Canada's trade surplus of $496 million with the British meant the country could not balance its trade books. Canadian holdings of United States dollars dropped to $1.6 billion by the end of 1946.[12] Less than a year later, in the fall of 1947, Canadian holdings of US dollars totalled $480 million.[13]

For Canadian spectators, the economic crisis of 1946–47 simply replayed the crisis of 1940–41. Again, Canada faced a massive shortage of available American dollars to cover its debts in the United States because of British currency policy. As a result, more than preliminary discussions about a free trade agreement between Canada and the United States took place in October 1947. During those meetings, Canadian emissaries proposed that a comprehensive agreement would include "wherever possible, the complete elimination of [customs] duties." However, Mackenzie King served as a brake to ongoing negotiations. Despite initial enthusiasm, the Prime Minister began to shy away from the formal free trade arrangement, fearing the political backlash a treaty might cause.[14]

Officials in Ottawa watched with growing alarm as Canada's trade deficit with the United States increased. Measures taken to combat inflation – like pegging the Canadian dollar to that of the United States – failed miserably and had the exact opposite effect of that intended by Canadian politicians. Instead of curbing inflation, Canadian consumers went on increased trips south of the border and increased Canada's dollar deficit $250 million to $1.041 billion. The situation became so dire that by late 1946 and early 1947, Canadian trade officials began drawing up lists of imported American "luxuries" that they could ban or sharply reduce.[15]

However, Canada relied on imports of strategic raw materials – like coal, oil, and iron – to such a degree that curtailing their consumption would slow the postwar economic boom. The development of indigenous natural resource capacity would help to sustain Canada's economic growth. The problem, however, was that making additions to Canada's crude oil reserves proved to be frustratingly elusive. Proven reserves from the country's largest field at Turner Valley were declining at a rapid rate. Production from Turner Valley field peaked in 1942 at just less than ten million barrels. By 1946, production totalled nearly six million barrels and continued to decline (Table 3.1).[16] "Despite aggressive exploration in the foothills and further searching on the plains," wrote David Breen, "no major fields had been found."[17] Imperial Oil, for example, spent approximately $23 million between 1917 and 1946 on exploration. Unfortunately, all they had to show for their effort was a string of 133 dry wells.[18]

On the other hand, little doubt remained about the potential reserve capacity of the oil sands. If successfully tapped, the sands could easily supply all of Canada's energy needs. The wartime experience with Abasand, however, soured Mackenzie King's government on directly pursuing

Table 3.1   Annual Oil Production from Turner Valley,
1945–60 (Barrels)

1945	7,009,521
1946	5,937,362
1947	5,022,350
1948	4,432,084
1949	3,826,543
1950	3,344,007
1951	2,952,302
1952	2,655,007
1953	2,404,967
1954	2,135,799
1955	2,054,840
1956	1,775,523
1957	1,594,220
1958	1,446,467
1959	1,338,404
1960	1,199,844

Source: Breen, *Alberta's Petroleum Industry*, p. 559.

development any further. Combined with King's unease about government intervention into the private sector during the war, the federal government was content to turn the matter back into private hands. However, there was deep concern over finding suitable investors. Reports published by the Department of Mines and Resources cast serious doubt that a solution to the oil sands riddle could be found.[19] Would future investors simply be wasting their money? More to the point, were there any investors left in Canada who would be willing to gamble that they would enjoy more success than Abasand Oils under Max Ball?

THE PROVINCIAL INITIATIVE

Fortunately for the province, another developer was waiting in the wings. Even before the end of the Second World War, a proposal by Lloyd Champion circulated in offices at the Legislature. Champion suggested that the province and his company launch a joint venture to continue development of the oil sands.[20] In the aftermath of the Abasand debacle, however, Champion was unable to raise enough private capital to finance a large-scale commercial development. With no other sources of capital readily available, Champion asked Edmonton to either join him or to refinance his company.[21]

The proposal represented a crossroads for the Alberta provincial government of Ernest Manning. On the one hand, his right-wing Social Credits believed there was little role for government in the private sector and suspiciously eyed attempts to enlarge government's role. During Manning's twenty-five years as premier of Alberta, provincial politics bucked national political trends by displaying cautious fiscal conservatism and equally cautious social reformism. On the other hand, Manning realized that if the government did not join forces with Champion, the oil sands might continue to sit undeveloped for the foreseeable future.

As Manning and his cabinet searched for answers, it seemed clear that there was little alternative. The province would direct the oil sands effort and serve as a catalyst for industry. Although Manning and his cabinet would have preferred to let the private sector – specifically major oil companies – develop the tar sands, the majors clearly had little intention of doing so. Karl Clark likened the oil companies' attitude to that of children trying new food; "they don't like it before they have tasted it." There was little doubt that the majors would begin development if the oil sands resembled conventional oil. However, because the oil sands were an unconventional fuel source, Manning realized that the government would have to blaze a trail for the private sector to follow. Confronted with a choice between doing nothing and doing something, Manning chose the latter. The province would enter a partnership with Lloyd Champion.

Having made the decision, Manning sought to minimize the risks the province would face. Manning asked Karl Clark to evaluate Champion's proposal and to recommend a course of action. Once again, Clark sifted through the evidence and combined it with information gleaned about the oil sands from his own research. Not surprisingly, Clark argued that solutions to the remaining technical problems of oil sands separation were near. It was critical that skills and techniques honed in the laboratory and small pilot-plant facilities found commercial application.[22] Without taking this next step, Clark feared that the separation methods perfected in the University of Alberta's labs could become another piece of academic inquiry with little "real world" application.

The key to further development did not lie in the hands of the provincial, or even federal, government. Rather, Clark firmly believed that it would only be with significant interest and investment by the private sector that the oil sands would realize their potential. However, he concurred with Manning's assessment that the province would have to act as a catalyst to the private sector. The key to attracting private investment would be a project undertaken by the provincial government to establish project costs. "If cost is not established soon," concluded Clark, "the resource may lie idle, needlessly, for years before some outside party … appears."[23]

Clark's assessment largely confirmed the conclusions contained in other independent reports solicited by the province. They offered few arguments to dissuade the government and momentum continued to build for the joint venture. In the rush to complete the agreement, the province left several loose ends. While the province agreed to provide Champion's Oil Sands Limited with $250,000, there were some indications that this investment would not see the plant through to completion. Opposition critics in the Alberta legislature openly questioned the wisdom of the venture because the deal struck by the Manning government seemed remarkably one-sided. The province would provide Oil Sands Limited with the financing to build the plant and provide a marketing board to procure materials for the plant's development. In exchange, Oil Sands Limited could pay back the provincial loan over the next ten years. Provincial taxpayers, argued critics, would bear all the risks and the company would reap the reward.[24]

Other commentators openly wondered if provincial politicians had learned anything from observing the federal effort during the Second World War. Provincial politicos seemed to replicate the same mistakes made by Ottawa.[25] Karl Clark, for one, assumed that a provincial/business joint effort would be free from the mistakes committed at Abasand. Surely, reasoned Clark, the province would consult the oil sands experts at the University of Alberta. The reality, however, was somewhat different. Oil Sands Limited was destined to replicate most of Abasand's failings while inventing a few new ones of their own.

Although the Trustees solicited the opinions of Clark and the Research Council, the consultations seemed perfunctory. Members of the Research Council grumbled that they exercised little influence on decision-making regarding even the most basic details of the new plant's operations. Finally, when the government hired a firm from Tulsa, Oklahoma to design the plant, the move stunned Research Council scientists. They pointed out that Born Engineering had no experience in the oil sands. To make matters worse, the company decided to continue with little participation from Research Council scientists. Perhaps blinded by their mistrust and suspicion of Ottawa, many key provincial decision-makers who attributed the Abasand disaster to "[federal] government ineptitude" believed they were immune from making similar mistakes.

Clark entertained no such illusions and frantically tried to bring Born Engineering up to speed when shipping delays halted work at the site in the late spring of 1945. In a letter to Dr. Sidney Born, Clark sought to give the firm a synopsis of working conditions in Athabasca just in case Dr. Born was "[the] sort of supposing that roses bloom at Bitumount at Xmas just as at Tulsa."[26] The work season at Athabasca was limited to six warm months between April and September. Since it was already late June, Clark tried to stress the importance of acting promptly. "Only 3½ months of the season remain. The implications re getting material landed at Bitumount this year is clear," wrote Clark. "There is not much room for a manufacturer to take six weeks to ship and then get it down river to the plant."[27]

Despite Clark's best efforts to spur the project along, spiralling costs and construction delays persisted through the remainder of 1945. "It took three months to buy the mill and deliver it," wrote a frustrated Clark. "If ordering and buying is not improved several thousand per cent the plant will not be completed in my lifetime."[28] Meanwhile, unease on the Board of Trustees prompted them to initiate cost-cutting measures much to the chagrin of those working directly in the camp. Work on the Bitumount plant lurched forward into its second summer with little progress made. In a letter to Sidney Blair in July 1946, Karl Clark confirmed that, despite the setbacks, the province would continue to invest in the oil sands. The "real motive force" behind the province's desire to bring the Bitumount plant online was the failed Abasand effort. "Our government pleases to regard the Ottawa government as the lowest thing on earth, incapable of doing anything right but capable of the lowest forms of political dirt such as deliberately sabotaging the chances of development of Alberta's great tar sand resource," wrote Clark.[29] "The last crack will not stick unless Alberta can produce a successful tar sand plant where Ottawa failed." The Abasand fiasco provided a useful club "to beat Ottawa with and while wielding the club, [Edmonton] appropriated $250,000 for a tar sand plant of their own" although "everyone" knew the appropriation would not be enough to see the project through to completion.[30]

Clark confessed that he initially regarded Champion as "the business brain and the boy who could roll all the politicians and others that needed rolling." In the interim, however, Champion "proved a washout so far as a good businessman is concerned." Champion's lack of managerial skills alienated Oil Sands Limited's first lead engineer and resulted in that man leaving the project. They found a replacement, but Clark clearly believed both men were merely competent engineers. Much to Clark's chagrin, Champion spent most of his time back east "doing I do not know what. However, one thing he is supposed to be doing is to find the extra money to supplement the government's contribution of $250,000. This he has failed to do, completely." Clark pointed out that the province had already agreed to boost the plant's capacity from 250 barrels per day to 350 barrels per day. The plant's expansion would cost an additional $400,000 to finance. Since Champion displayed little success at raising outside capital, Clark feared that the province would be forced to make another appropriation, requiring time-consuming debate in the Alberta Legislature.[31]

To be certain, many problems that conspired to delay completion of the project were the faults neither of the province nor of Lloyd Champion. Postwar shortages of materiel and skilled tradesmen delayed construction and inflation drove the price of materials higher. While he anticipated some factors hampering development, a sense of urgency gripped Clark. In the aftermath of the Second World War, the oil sands had a brief window of opportunity to emerge as a viable primary fuel source. Postwar shortages of fuel oil and gasoline spiked world prices. Perhaps more important, world crude supply was not enough to meet consumer demand. Producing countries, like the United States, began to consider setting up quotas on exports to conserve supply.[32]

SYNTHETICS ASCENDANT?

The changes to the world market were temporary. However, a series of factors conspired against the emergence of a market at precisely the same time the United States government began its own synthetic fuel investigation. Subsumed within America's role as Cold War leader were national security interests that included the articulation and development of a comprehensive oil strategy. In the era of "permanent crisis," the United States National Security Resources Board (NSRB) was one mechanism available to achieve maximum preparedness.[33]

Secretary of the Interior, Julius Krug, White House assistant Oscar Chapman, assistant to the President John Steelman, and many other officials believed that the national interest required a commercial synfuel plant well beyond the scale of a demonstration plant. Krug and the Secretary of War, Robert Patterson, envisioned a synthetic industry capable of producing a million barrels of fuel per day to preserve national security. However, not only did oil sustain technologically advanced and equipped military forces, it also drove the industrial machinery that formed the backbone of the West's peacetime economy. Quite simply, the absence of war did not prevent the need for secure and reliable supplies of oil. Thus, policymakers framed the discussion of a postwar petroleum strategy around two crucial questions:

To what extent American oil reserves should be conserved by imports from areas beyond the American sphere (i.e., the Middle East) while the international situation permits, and (2) To what extent the oil needs of the countries of Western Europe, now supplied largely from the Middle East, are related to those of the United States and to what extent they have become a part of our national responsibility. [34]

The Joint Chiefs of Staff weighed into the debate and warned that in a future conflict, the United States could run out of oil. Although the oilfields of the Middle East could produce a large quantity of oil for the least amount of investment, the Chiefs acknowledged that the region was "very susceptible to enemy interference." More important, the Middle East was a "poor risk from the standpoint of accessibility of production in case of war." [35]

The Chiefs suggested that the United States adopt a three-point program that would guarantee the country's energy future. America should begin developing its reserves of natural gas, coal, and oil shale. Meanwhile, the government should to begin the process of research and development that would make a commercial synthetic fuels industry more viable. The United States should also court countries with proven reserves, particularly in the Western Hemisphere and the East Indies. These sources could supply conventional crude to the industries and markets of the United States and its allies. Finally, the United States should conserve domestic crude through the "maximum importation of Middle East oil." [36] Thus, the Cold War meant that conventional crude supplies from the Middle East were now a matter of national and international importance.

Significantly, the Joint Chiefs argued that the development of a synthetic fuel industry offered the United States a strategic reserve that would be relatively immune from enemy action. Synthetic fuels, they argued, provide "the most promise as the safest and most prolific means to provide the required supply of petroleum and petroleum products to meet the demands of a future war." [37] Unfortunately, the American program never bore fruit for the same reasons that the Bitumount plant in Alberta foundered. In the 1940s, the Bureau of Mines constructed and operated several experimental synthetic fuel plants in several different locations throughout the country.

The pilot projects, however, were never able to make the transition from experimental to commercial operations.[38]

In addition to being capital-intensive projects, synthetic fuels also required larger amounts of steel and other raw materials compared to the production of conventional oil. Analysts regarded the situation warily. "It has been estimated that 49 tons of steel are required for plant construction per daily ton of liquid fuel derived from natural gas; 70 tons of steel per daily ton of oil from coal; and 34 tons of steel per daily ton of low-grade oil from shales and additional amounts for the further refining of crude shale oils," concluded one report. "In contrast, a modern refinery for crude petroleum calls for about 6 tons of steel per daily ton of oil products."[39] Thus, the program began on shaky ground long before conventional producers could object to synthetic crude displacing conventional sources on the world market.[40]

In retrospect, several factors contributed to the shelving of the synthetic fuels program. Expected technological advances in the 1940s and early 1950s to make the production of oil sands and oil shale economically feasible did occur. However, they also had to compete with advances in other areas, particularly in the field of natural gas, and could not keep pace. Before the Second World War, natural gas was little more than an unwanted by-product that drillers flared from the top of wells. After 1945, however, natural gas gradually emerged as an alternative energy source. With advances in pipeline technology, particularly with respect to electric welding and high-pressure equipment, the main impediments to the construction of large-scale pipelines disappeared. As a result, natural gas quickly replaced coal as a major fuel source, eliminating the need for one-twelfth of coal's annual production. In the fifteen years between 1940 and 1955, production of natural gas in the United States alone rose from 2,660.2 billion cubic feet to 9,405.3 billion cubic feet. Meanwhile, between 1930 and 1960, commercial and private consumption increased more than ten times.[41]

Besides increased North American energy supplies, synthetic fuels also found themselves competing for market shares against prolific global suppliers of conventional crude. American interest and participation in the Middle East increased, due, in no small part, to the emerging Cold War and the vast supplies of oil in the region.[42] With the power of the British on the wane, it fell to the United States to secure the region and its petroleum. Oil became a strategic commodity vital to the maintenance of America's role as leader of the Western alliance. The Middle East was important for a variety of reasons, most significant of which were the vast oil fields.

In the global game, the United States effectively won two victories over the Soviet Union by incorporating the Middle East into its sphere of influence. Not only did the Americans gain access to the largest deposits of conventional crude in the world, but they also denied those very same resources to the Soviet Union. Furthermore, American-based multinationals derived significant income from their concessions in the region. The State Department widely assumed that American investment abroad, particularly in oil, gave the United States added prestige and influence in the region.

Perhaps the lessons of the 1930s weighed heavily on the minds of American policymakers. They could not permit Soviet expansion and acquisition of raw materials lest the Soviets use them against the United States. If Britain and the United States "lost" Iran, it would cost approximately $700 million annually to find alternative fuel sources to supply Western Europe.[43] "The military and economic costs of regaining even a portion of these resources," argued the National Security Council, "should they become essential, would be enormous. The military and diplomatic measures needed to hold that area, should its oil be vital to the entire war effort, would be most difficult to implement and uncertain to succeed."[44] Moreover, as Melvyn Leffler concluded later, "if [Stalin] tried to exploit prevailing circumstances, Soviet Russia might gradually co-opt, either directly or indirectly, enormous industrial infrastructure, natural resources, and skilled labour," directly threatening America's national security.[45] Thus, to paraphrase Daniel Yergin, in the postwar era, access to Middle Eastern oil was the point at which diplomacy, foreign policy, international economic considerations, national security, and corporate interests all converged.[46]

LEDUC AND THE OIL BOOM

The combination of these separate, but nonetheless related, developments placed the future of the oil sands in jeopardy. Without a market to service, would a private investor be willing to spend the kind of money required to build a commercial oil sands plant? Nevertheless, construction at the Bitumount plant lurched toward completion in early 1947 when Imperial finally struck oil at Leduc # 1 just south of Edmonton. At the end of 1947, thirty different wells pumped 3,500 barrels per day from the field. "The real oil boom that a generation of Albertans had dreamed of was under way," wrote David Breen. "Leduc roused the interest of international capital in Alberta's potential, and long-awaited development funds poured into the province."[47]

Discoveries of large oilfields at Leduc and a year later at Redwater radically altered the Province of Alberta's commitment to the oil sands. The two conventional fields contained a total of 900 million barrels of oil, more than the estimated 800 million barrels accessible from the oil sands given proven mining techniques. Furthermore, increased royalty revenues from conventional wells eliminated whatever economic impetus there may have been to development. The increase in conventional production to 45.9 million barrels – approximately 57 per cent of Canadian consumption – ensured the oil sands' status as a strategic reserve. The emergence of a significant conventional oil and gas industry jeopardized the future of the oil sands. In 1949, production from Leduc already exceeded the demand in the only market available to Alberta crude – the Prairie Provinces themselves.[48]

Combined with the construction delays at Bitumount and the precarious economic condition of Lloyd Champion's Oil Sands Limited, the future of the oil sands looked grim. Politicians and businessmen now had the luxury of regarding the sands as simply another resource in Alberta's quiver. Although the Manning government did not totally abandon research on the oil sands, the sense of urgency that guided earlier efforts noticeably diminished. The sands now assumed the dubious distinction of forming the core of Alberta's "second line" of energy resources.

For Canadian producers, the emerging emphasis on synthetic fuels as a strategic reserve was a mixed blessing. Edmonton and Ottawa had sponsored research and development on the oil sands since the 1920s. To be sure, they wanted to see some return on their investment, but a large bolt of pragmatism informed provincial decision-making. There was little guarantee that the United States would be willing to purchase synthetic fuel from

Canada while cheap conventional oil was readily available. Furthermore, if the growth and development of the oil sands continued, Canada's own conventional industry would suffer and the province would lose the valuable royalties it collected from producers. Without an external market to sell its oil, the province would not garner significant revenues – already a multimillion dollar source for the province. During Alberta's first decade of significant oil production, provincial revenues increased from $68.1 million in 1948 to $289.2 million in 1957.[49]

Nevertheless, the province had expended too much time and treasure on oil sands investigations to turn back now. Clark and the Alberta Research Council pleaded with the province to continue funding oil sands research and to complete construction of the Bitumount plant. The oil sands needed money and research to determine the commercial viability of the unconventional fuel source for the marketplace. Without funding and research, the province would not know whether the oil sands were "a second line of defence or any defence at all."[50] The province agreed with Clark to a certain extent, and continued funding research on the oil sands in part because of its value as a strategic reserve.

It is important to remember that the oil sands were one small corner of the North American energy market.[51] In fact, the true "second line" of defence was North America's shut-in capacity and not its unconventional fuel sources. The American oil industry enjoyed a banner year in 1948 as petroleum stocks increased by as much as 20 per cent. The Texas Railroad Commission reduced American output from 2,710,000 barrels per day in December 1948 to 1,872,000 barrels per day in July 1949.[52] The industry estimated that with proven reserves at current rates of production, there was enough oil "in the ground" to meet demand for the next nineteen years. Yet over the same period, proven world oil reserves increased from sixty-two billion barrels to 534 billion barrels in 1972.[53] Thus, just as fast as the world consumed oil, new conventional reserves replaced depleted supplies.

Foreign imports, specifically from the Middle East and Venezuela, continued to flow into North America and made up another prong in the strategic reserve, albeit for a vastly different reason. Conservationists had long argued that the United States should conserve its own deposits of conventional crude for as long as possible. At the forefront of this endeavour was Charles Leith. A geologist by training at the University of Wisconsin, Leith floated in and out of government since the turn of the century. As a technical adviser to Woodrow Wilson at Paris in 1919, Leith quickly developed a hearty disdain for the professional diplomats in charge of carrying out the nation's policies.[54]

Leith argued that American minerals policy should implement a variation of David Ricardo's theory of comparative advantage. A liberal world order would be the most efficient and equitable means to allocate mineral production. Increased specialization would mean increased prosperity for all nations in the newly emerging global system. Thus, at the end of the Second World War, Leith argued against more domestic production and instead suggested that the government stockpile foreign supplies. Purchasing vast quantities of raw materials and resources from abroad would stimulate local economies and would put much needed capital back into the system.[55] Fuel imports also delayed the deterioration of North America's conventional reserves. Therefore, this policy indefinitely pushed back the day when unconventional fuel sources could expect to capture a significant portion of the North American energy market.

The shut-in capacity of North American petroleum and the ready availability of cheap imported crude meant there was too much oil on the world market. The multitude of energy sources flooded the market and drove prices down. In response to falling prices, the oil majors began cancelling their own synthetic fuel experiments voluntarily. Standard of New Jersey stopped working on a coal-hydrogenation pilot plant. Standard of Indiana dropped its plans to build an $80 million natural gas-hydrogenation plant, and Union Oil stopped all work on oil shale.[56] With their own projects now on the shelves, the industry turned to the government and insisted that they divest themselves of their synthetic fuel programs as well. The oil industry "vigorously opposed" the federal government's continuing support of the synthetic fuels program.[57]

Domestic producers objected to the development of new competition at public expense. Foreign concessions were under pressure from host governments to increase production. Moreover, oil imports were highly profitable, at least if domestic pro-rationing bodies maintained an artificially high price in the United States. These efforts were ultimately successful as the Eisenhower administration cancelled Truman's synthetic fuels program in 1954.[58]

At roughly the same time that American firms re-evaluated their commitment to synthetic fuels, the Province of Alberta began to receive the first hard data from Bitumount. The pilot plant had a difficult year in 1948. Lloyd Champion was unable to honour his portion of the agreement with the province and disappeared from the project altogether. The provincial government ran Bitumount's operations. Meanwhile, a fire destroyed the machine shop and warehouse in May, all of which needed rebuilding before operations could begin that summer. Despite these setbacks, and the inevitable start-up delays, tests continued for a few restricted hours when the entire facility was ready in late August. Researchers discovered a few problems with the plant's set-up and design, and moved to correct them. Despite the many obstacles encountered – never during the season was the entire plant operated simultaneously – the Research Council declared itself pleased with the facility and the research it enabled. "The outcome of the season at Bitumount constitutes a distinct advance beyond what has been accomplished heretofore in bituminous sands separation plant design and operation," crowed the writers of the Research Council's Annual Report.[59] "While it is true that the whole plant was not run simultaneously, it was only the lack of operators that prevented this from being done and continued indefinitely."[60]

However, as Karl Clark confided in a letter to Sidney Ells in early November, there were still reasons for concern. The Research Council was not officially involved with the project; while the province expected the Council to offer advice and to be helpful, it exercised no real authority or power. Much to Clark's displeasure, the Council could only suggest changes during the planning stages and correct problems afterwards, further delaying the time when the plant would be operational. "Every plant that has been put together to date has made oil, " wrote Clark, "the difficulty is that the engineers have not yet learned how to design a plant for giving expression to the hot water process which will keep running without getting into mechanical difficulties."[61]

Although he made several suggestions to help improve the plant's design, Clark often found his suggestions ignored by the engineers the firm hired to construct the plant. "I am told that these are matters of engineering and that I have no particular competence in this field," complained Clark. "Well, they probably do but they do not know enough and certainly there is no common basis of knowledge from extensive experience in building separation plants." With little alternative available, Clark concluded "we will have to take what the engineers turned out for a start and modify it locally till it works."[62]

The Board of Trustees believed they had to convince the Alberta government of the need to continue research and development. To this end, the Trustees tabled a report in the Provincial Legislature in the spring of 1949 and provided the first really systematic account of the Bitumount plant since its inception. The report began by defending the need to continue research and development of the oil sands and hinted at the need for a national energy policy.

Canada, the authors maintained, was dependent on foreign oil, and domestic demand continued to grow. While acknowledging Alberta's emerging conventional industry, the Trustees argued that the oil sands project represented a major source of domestic oil. The pilot plant at Bitumount was an essential investment in the future. Only by proving the viability of the oil sands could Alberta expect private investors to put their money into the development of this vital resource. The report did not sugarcoat the many operating difficulties encountered at Bitumount during its brief operating season the year before. Instead, the report's authors concentrated on the problems identified and solutions started before the end of the season. The province could rest assured that Bitumount would not be the massive failure Abasand had been.[63]

Reassured somewhat by the findings, the province continued its support of the Bitumount plant for another year. However, Edmonton overlooked suggestions that the oil sands replace imported fuel and become a cornerstone of a national energy strategy. In the first place, conventional oil accounted for most provincial revenues and the province produced more oil than all the Prairie Provinces could consume. This occurred despite a provincial ban on export of oil products to other parts of Canada or the United States until Alberta could guarantee itself a thirty-year supply of oil.[64] Clearly, introducing more production onto the market would depress prices even further.

Secondly, although the province regulated oil and gas production, the articulation of a national energy strategy required the involvement of the federal government. Given the direct and indirect costs involved with carrying out such a program, Edmonton did not expect, nor did it receive, much encouragement from Ottawa. Instead, the province hinted that the 1949 season would be the time for full-fledged tests to take place. The plant had another year to prove that the oil sands were a resource suitable for commercial development.

The plant did quite well after revisions and upgrades to plant equipment were completed in 1949. "It was evident that when proper adjustments of conditions were made [the plant] would give gratifying results," noted Clark.[65] Although a series of mishaps and explosions prevented the facility from operating for part of the season, the plant generated enough data to deem the project a success. In late September, Sidney Blair, who had previously worked on the oil sands for the Research Council in the 1920s, and Ed Nelson, the vice-president of Universal Oil Products Company, visited the Bitumount site. The province asked the two oilmen to provide their opinion about the future of the pilot plant. Should the province fund another year of research, or was it time to move on to the next phase of operations? With Clark serving as their guide, the two guests toured the facility and inspected the data collected by the Research Council.[66]

During their visit, it was clear to both men that there was little more to gain from continuing the plant's operations. Bitumount had served its purpose and showed that there was no technical obstacle to the separation of oil from the sands. Blair and Nelson both impressed on Clark that the next step was an estimate of the costs of production for a barrel of commercial oil sands oil. Nelson amplified the point in a subsequent exchange of letters with the province.[67] "Such an analysis would reveal the weak spots in all that was necessary for a successful business," agreed Clark in a letter to his son. "I am still rather mystified about how one gets going on this broad analysis but maybe I shall see light."[68]

Clark need not have worried about preparing the report himself. Instead, the province wanted an outside party, with experience in the oil industry, to write the study and to inaugurate the next phase of oil sands development. Under Ernest Manning's guidance, the province took development of the resource as far as it could on its own. Decades of research and development culminated in the successful operation of the Bitumount pilot plant and now it was time to begin commercial development of the resource.

CHAPTER 4

From Scientific Project to Commercial Endeavour

May the words "welcome" and "oil sands" ever be closely associated! – *Honourable Gordon E. Taylor, Minister of Highways, Government of Alberta, September 1951*[1]

Growth in the Alberta oil patch between 1949 and 1951 offered significant promise for the province's economy. Supercharged by the influx of petroleum revenues, royalties, and associated taxes, and with more oil and gas sources coming on stream, the future seemed limitless. However, by the mid-1950s, Alberta's oil industry was in a state of crisis. With nearly half its productive capacity "shut-in," the market for Canadian crude evaporated. Faced with the uncomfortable choice of abandoning oil sands development or alienating the province's conventional producers, the Manning government coolly assessed the situation. With no fanfare, Edmonton quietly decided to place the oil sands on the back burner through a series of shrewdly plotted manoeuvres. Abandoning the usual political rhetoric, they simply created a set of conditions that made commercial development of the oil sands exceedingly unlikely. The province implemented leasing conditions and marketing possibilities that braked business enthusiasm. The new provisions ensured that Alberta's conventional oil and gas would not lose its share of the North American market.

At the beginning of the 1950s, a sense of optimism prevailed among oil sands boosters. The sands gradually altered their status from a scientific project to a matter for government policy. In early 1949, it was clear to the government of Premier Ernest Manning that the province was facing critical choices. With the large finds of conventional crude in the province, the driving forces behind tar sands research – the need to become energy self-sufficient – waned appreciably. However, several provincial governments, including Manning's, had invested millions of dollars over the decades to unlock the riddle of the oil sands. It did not make sense to abandon the project now. Thus, Manning found himself on the horns of a dilemma. What should the province do with the oil sands?

The province's best and most knowledgeable experts all believed that they were on the verge of a major discovery. Karl Clark's persistent investigation and improvements to the hot-water separation encouraged Research Council scientists. Perhaps, with a little more time and money, the province could finally add the oil sands to the list of producing oil fields in Alberta. Nevertheless, it was also clear to many observers, including Clark and the Research Council, that the pilot plant at Bitumount had outlived its usefulness. Research Council scientists could realistically expect to squeeze little more information out of the facility than they had already gathered. While they had largely settled the technical questions surrounding oil sands development, many problems remained. If a barrel of oil sands oil could not compete with a barrel of conventional crude on the world market, little point remained in continuing investigations.

However, to determine whether the process was economically feasible, the Research Council argued they needed to upgrade their facilities. That, in turn, would require another significant investment by the province. This was controversial. Opposition members in the legislature questioned the need for the province to sponsor operations like the Bitumount plant. At a relatively innocuous third reading on an appropriation bill during a March 1949 session, the experimental plant at Bitumount drew the ire of Opposition members. After listening to Opposition complaints, the Minister of Industries and Labour, Dr. John L. Robinson, rose to answer their questions. He acknowledged that he was uncertain whether the plant was a "lemon" or a "plum" but urged the legislators to appropriate the requested funds regardless. Unfortunately, no experiment could guarantee commercial applicability but the province had to press on.[2]

Although the appropriation passed, the question struck at the core of the government's dilemma. Was the oil sands a viable industry or was the government simply throwing good money after bad? The emergence of Bitumount as a political issue convinced the provincial government that the next logical step would be to figure out the sands' cost of production.

THE BLAIR REPORT

After the visit by Nelson and Blair in September 1949, the province asked the two oil men to assess the known mining and separation methods. The government wanted them to provide estimated production costs for a barrel of saleable oil. The Department of the Industries and Labour, under the direction of Minister John L. Robinson, supervised the report. Ideally, the project would summarize the data collected by the Alberta Research Council and encourage private development of the oil sands. Until the province made the estimate "there could be no answer to the question of whether oil sand development was a matter of present day concern," concluded Clark. "The attitude of the oil industry in the existing state of knowledge was that the oil sands were very much in the future."[3]

While Blair was organizing data for his report, a memorandum arrived from Karl Clark outlining his thoughts on the future of the oil sands. "The common talk is that discovery of prolific oilfields in Alberta eliminates the bituminous sands from further consideration," wrote Clark. "It is more logical to take the opposite view." Rather than limiting oil sands development, Clark argued that the province should vigorously pursue development and increase Alberta's available reserves. By increasing the province's capacity, the provincial government and the oil companies would surely see the need to build the necessary infrastructure to transport its goods to market. "When the economics become right for bituminous sand oil, transportation will be a minor instead of a major part of the whole development project." With the infrastructure in place, it would simply be a matter of making a minor connection to the existing main pipeline. Clark summarily dismissed arguments that an overall increase in Alberta's production would depress world crude prices. "It is not likely that Alberta oilfields will glut the Canadian, let alone the North American, market in the years ahead."[4]

Although Clark correctly assessed Alberta's role in the North American energy market, his enthusiasm and dedication for oil sands development blinded him to the realities of the world petroleum market. The presence of excess capacity did not necessarily bring with it assurances that further development would take place. While Clark was correct to say that Alberta did not produce enough oil to dominate the North American market, he did not account for other factors inhibiting development. Most important was the fact that a barrel of synthetic oil could not compete with conventional supplies on the open market. Nothing, besides its higher production costs, distinguished it from conventional crude. Furthermore, considering that Alberta's oil served a limited market on the Canadian prairies and select areas of the Western United States, conventional producers were more than able to meet demand. In the 1950s Alberta oil gradually made its way into the Great Lakes areas of Ontario and the United States. Eventually, in 1954, the province would export small amounts to Washington State and California, but Canada's modest share of these markets did not warrant the introduction of a new fuel source.[5]

Despite the comparatively small volume of exports, the province was in danger of saturating its market and driving down the price conventional producers could expect. The discovery of the large Redwater field in 1948 and the discovery in 1949 of four additional reservoirs (Joseph Lake, Golden Spike, Settler, and Excelsior) contributed to the province's dilemma. Beginning in 1949, Alberta's conventional production potential exceeded market demand. To protect their assets, industry representatives then approached the provincial government about beginning a pro-rationing system.

After conducting public hearings, the province acted on the request of the conventional industry. Edmonton introduced a pro-rationing system, monitored and administered by the Conservation Board, to ensure that the production of Alberta's oil did not drive down the world price for oil. Originally created by the Province of Alberta in 1938, the original purpose of the Conservation Board was to ensure that proper procedures were followed in the development of Alberta's oil and gas resources. The Board's mandate now expanded to include the regulation of production to ensure all producers equitable access to markets. After examining data from each oil field in the province to learn its producing potential, the Board would determine the "maximum permissible rate" (MPR) – how much oil per month a given field could produce for sale on the world market – and regulate production accordingly. Thus, with each additional conventional oil field added to Alberta's inventory, the amount of oil produced by the province's other fields shrank proportionally. The spectre of adding the oil sands output to this total loomed menacingly on the horizon for the province's conventional producers.[6]

However, with each conventional well drilled, the focus of provincial politicians gradually changed. Simple economic realities dictated that the province must continue its support of conventional oil and gas producers. Royalty payments from oil and gas producers gave the provincial government the highest per capita revenue in the country. In 1955–56, for example, Alberta took in $225 per resident in taxes and royalties compared with a national average of $125 per resident.[7] Instead of needing to develop commercial synthetic fuels, the province now realized its first objective was to market the conventional fuel it already possessed.

Further undercutting Clark's argument was the technological state of the synthetic fuel industry itself. Although the Athabasca deposits literally contained billions of barrels of raw bitumen, the most optimistic assessment claimed that only 700 million barrels of bitumen were recoverable given current technology. Conversely, the two largest conventional oil fields in the province, Leduc and Redwater, contained 200 million and 700 million barrels of oil respectively. Since the conventional industry in the province had greater proven reserves, conventional producers exercised tremendous influence in determining how the province handled matters affecting the industry.[8]

Nevertheless, Blair's mission was to assess the advantages and liabilities of all the alternatives available to synthetic producers. His second task would be to identify the best sequence of operations that would maximize both profits and production. "The choice would be governed by the technical suitability of operations, the way they fitted together and their costs," summarized the Alberta Research Council in its yearly report. From this, the province wanted to figure out whether the development of the oil sands "would be profitable now or in the immediate future." They also hoped that the Blair report would identify "weak spots" in the sequence of operations that stood in the way of development. Once these "weak spots" were determined, they could develop and prioritize research programs. Finally, the government intended to use the report to decide the usefulness of the Bitumount plant and site for further research.[9]

Almost immediately, Blair encountered a formidable obstacle. As he began synthesizing the Bitumount data, Blair found that the facility did not generate all the required information to write the report. W.E. Adkins, the supervisor of the Bitumount plant during the course of its operations, handled all the correspondence between Blair and the facility. It fell to Adkins to inform Blair about the paucity of data. "There are some points on which no information is available although it might be possible to make some fairly accurate estimates in these cases," wrote Adkins.[10] Along with the request for hard data, Blair asked Adkins to provide a detailed description of the site's geological qualities; that is "any stratification in the quarry, an analysis of the bitumen and of the sand, mentioning any variations in quality that are sufficient to effect operating costs."[11]

However, Bitumount did not undertake a significant mining program. As Adkins admitted to the Alberta Research Council, "Our knowledge of the overall characteristics of the oil sand at this point is very limited." Nevertheless, Adkins did not totally abandon Blair's request, and approached the Alberta Research Council about obtaining some of Karl Clark's field notes. In 1945, Dr. Clark dug some test pits and conducted studies on the bank immediately below the open pit mine. "We assume that he kept records of his observations at that time and we would suggest that this data now constitutes the only source of information on the general character of the sand in our mine."[12]

Understandably alarmed by Adkins' March 10 letter, Blair replied that the lack of information coming from Bitumount was cause for great concern. In his initial request, Blair outlined "what we consider absolute minimum requirements from pilot plants for determining the economic or engineering merit of the process for which a pilot plant has been set up." Without this information, preparing a report with any degree of accuracy would be impossible. Perhaps, he suggested, rather than lacking the appropriate data, Adkins simply misinterpreted what Blair wanted "due to the wording of the letters." In those instances where Adkins wrote that the plant personnel did not gather or keep detailed records, Blair simply suggested that they rely on "average" conditions calculated from spot tests.[13] Reassured and redirected by Blair, the first major crisis of the report's preparation passed as a minor hiccup. Adkins and the crew at Bitumount now understood what Blair required and worked steadily to give him all the information they could.[14]

The only other major obstacle Blair encountered during the report's preparation was the number of inquiries from interested third parties who hoped to glean some advance information about his conclusions.[15] Blair carefully guarded all material for two very important reasons. The first, and most obvious, is that he worked for the Province of Alberta and his first obligation was to satisfy the needs of his client. The second, and equally important reason was his belief that past work on the oil sands suffered because of "spotted or sectional attacks on the problem"; that is, for example, supporters of one extraction method undermining the efforts of others working on a different process. As Blair emphasized in a letter to John L. Robinson in mid-October, 1950, the contents and conclusions of the report must remain secret until its official release. Leaks could not only tip the government's hand, but Blair also argued they could fatally undermine the oil sands effort if critics and biased observers had ammunition to nitpick. "If single phases were considered separately from a Survey such as this, even by qualified people, they would be very misleading."[16]

Despite his obvious enthusiasm for the oil sands, Blair laboured under few illusions. Perhaps more than any other participant in this endeavour, Blair understood that there was little incentive for oil and gas companies to begin such a capital-intensive project. Why should companies invest millions of dollars in research projects when production of conventional oil was so inexpensive? Still, Blair did not want to deal the oil sands a death blow by denying that conditions might appear one day that would make development feasible. From this perspective then, the Blair Report provided the Province of Alberta exactly what it needed. It suggested that the oil sands could become a profitable venture but stopped well short of saying that its importance would surpass that of the conventional industry.

In early December, Blair presented his report to the province and re-leased the findings to the media. Blair's most important conclusion was that production costs totalled $3.10 per barrel but that the product had a market value of $3.50 per barrel. Clearly, argued Blair, the oil sands were "entering the stage of possible commercial development."[17] Less publicized, however, was the fact that Blair's report drew an explicit link between the size of an oil sands plant and the operation's overall profitability. "The capacity of any development must be relatively high in order to secure efficient transporta-tion and handling costs," concluded Blair.[18]

Reflecting the importance of Blair's findings, the provincial media greet-ed the official release of the report with great fanfare. "Exploitation of Oil Sands Said Feasible" cried the two-inch banner headline of *The Edmonton Bulletin*.[19] Others immediately grasped the importance of the presence of a vast, secure source of petroleum in the western hemisphere. The *Calgary Herald's* Andrew Snaddon immediately hailed the oil sands as "a vital cog in the democratic arsenal if war cuts off other oil supplies."[20] The warm recep-tion, however, did little to alter Blair's opinion that development of the oil sands was possible, rather than probable. Simply stated, the oil sands were an expensive proposition for any producer to undertake with no guaranteed market available to it. If the development of the sands were to continue, it increasingly looked as though it would be exclusively for an export market. This prospect was not appealing to Alberta's politicians because of public opposition to exporting too much oil.[21]

In May, John Robinson floated a trial balloon during a Rotary luncheon in Calgary where he suggested that "U.S. markets" could absorb production from the oil sands.[22] Given the importance petroleum supplies assumed after the outbreak of the Korean War in July 1950, Robinson was clearly looking to capture the markets of the Pacific Northwest. If the United States would be willing to open the Pacific Northwest and California to additional Alberta crude, the Canadian oil patch would be on solid foot-ing. The industry could expand operations, possibly even bringing the oil sands on stream much earlier than anticipated. Although the United States government helped to nurture the Canadian industry in its infancy, it would not go as far as Robinson hoped. Canada's conventional oil producers still could not enter the lucrative West Coast markets.

Nonetheless, the Province moved to capitalize on the Blair Report's positive image by making the Research Council's technical data readily available to the oil industry. To help simplify the process, Blair suggested that the province should sponsor a symposium on the oil sands within a year of the report's release. Now that they published the report, the provincial government literally received hundreds of requests for copies. Moreover, the planned oil sands conference was attracting a great deal of interest. Clearly, the report stimulated interest in the oil sands. A delegation from the Anglo-Iranian Oil Company (the forerunner of British Petroleum), even came to Canada to evaluate the situation first hand. "We have come to see if the Blair report can be believed," said Dr. D.A. Howes, the head of Anglo-Iranian's research and development division, upon his arrival in Edmonton.[23]

Anglo-Iranian's interest in the oil sands excited the provincial media as they covered the delegation's every move. "Whatever doubts may have lingered in any Alberta minds, on the significance of the Petroleum Engineer Blair's report on the McMurray oil sands, must have been stilled yesterday by the announcement that one of the world's great oil companies is interested," opined the editorial writers at *The Edmonton Journal*. "When the Anglo-Iranian Company sends a group of its experts to 'take a look' at the possibilities, importance of the McMurray deposits to the world, and to Alberta is immediately conceded."[24]

Howes remained tight-lipped and declined to comment when asked for his impressions after Clark and Blair provided a personal tour of the oil sands. A stream of top-level executives from other companies also travelled to Alberta to investigate the deposits for themselves. When reporters asked the president of Standard Oil of New Jersey to comment on the oil sands in late May, he politely declined. Instead, an unidentified vice-president at Imperial Oil more familiar with operations in Northern Alberta issued a statement to the press. The oil sands were "an awfully hard way to make money" given the current state of technology. This did not preclude the possibility that improved technology would change the situation. In any case, Standard and its Canadian subsidiary made it clear they were more than happy with their conventional fields. There were no serious plans to go into the oil sands.[25]

For the remainder of the spring and summer months of 1951, the Manning government focused on preparations for the symposium. Then, in early August, the province received a devastating blow. John Robinson revealed to Blair that Anglo-Iranian's head of R&D, D.A. Howes, had expressed scepticism about the Blair Report's basic conclusions. In a 164-page assessment for Anglo-Iranian, Howes cast significant doubt on the official findings of the provincial government and detailed the areas where he thought Blair's assumptions were faulty. Blair's estimates, wrote Howes, "[ignore] interest charges on capital, and on-stream efficiency for the process units. The allowance for offsite facilities etc., is only about 40% of the cost of the process units." Perhaps the most damaging of Howes' charges was that Blair vastly overestimated the market value of the product. Rather than yielding a profit of $0.40 per barrel, Howes estimated that oil from the sands would cost producers an additional $0.74 per barrel. Instead of fetching the $3.50 per barrel, Blair claimed, Howes estimated that the synthetic crude would only be worth $3.06 in the Great Lakes market.[26]

In any case, Howes' assessment touched on the most troublesome point when he suggested that production from the oil sands did not fill a new market. Alberta's conventional production met approximately one-third of Canada's demand and cheap foreign imports supplied the difference. Any additional fuel supplied by the oil sands would have to bump one of these existing two sources. It was extremely unlikely that Eastern Canadian consumers would give up their cheap supply for more costly Alberta crude, let alone even more expensive oil sands. Thus, the provincial government faced an unappealing scenario. The tar sands would either encroach on the markets of the conventional industry or displace the cheap foreign imports that supplied Eastern markets. If the former occurred, the Province of Alberta could expect lower revenues. If the latter, Edmonton would be responsible for increasing fuel prices paid by Eastern consumers. Neither option was particularly appealing.

Anglo-Iranian's negative report poured a bucket of cold water on provincial euphoria. Privately, Robinson indicated that the government would stand behind Blair's evaluation of the hot-water separation process. Nevertheless, Robinson realized that a commercial separation facility had several other associated costs that dramatically reduced profitability. In any case, the most important point for the government was to continue preparations for the conference in September. "Neither this office nor the Department of Economic Affairs is making any mention of the contents of this Report," wrote Robinson. "I am suggesting that nothing be said about it until at least our Symposium is over with."[27] Clearly, the province intended to follow through with its plans for the time being. With so little time left before the beginning of the conference, there was little alternative.

Less than a month later, the province's first oil sands conference began. Senior oil company executives, officials from various levels of government, and a score of scientists gathered at the University of Alberta to hear and discuss papers on a variety of oil sands-related topics. The conference also gave the province an opportunity to outline both its leasing and royalty strategies. Nathan Tanner, the Minister of Mines, emphasized that the Province would do all it could to encourage development of the oil sands. This endeavour was "in the interest of the people of the Province and of Canada as a whole, and, further, to the security of this continent."[28]

Any company or individual who wanted to undertake a geological or geophysical exploration of the oil sands area would have to obtain a license from the Department of Mines. Should the company then want to take out a lease on a given parcel of land, they would then have to apply for a twenty-one-year lease. Rent would be $1.00 per acre and the province would also collect a royalty of not more than 10 per cent "of the products extracted from the sands."[29] The lessee had one year to begin building a commercial plant. In concluding his remarks, Tanner emphasized the importance of the oil sands for the security of the North American continent. The government's leasing strategy would "encourage immediate development" of the oil sands and *further the security of this continent.*" Lest anyone miss his final point, Tanner reminded his audience "it is important that we carry out a good program of development of our natural resources in the interest of security."[30]

For Tanner, the development of the oil sands was important not merely because of its significance as a fuel source. Indeed, the oil sands represented a test from Heaven to ensure that a "Christian way of life" survived. Tanner exhorted the assembled audience to join with him in this battle where the commercial development of Alberta's oil sands was tantamount to a crusade.

He reminded his listeners that the development of public policy was a responsibility the Manning government took seriously, particularly as it related to natural resource issues.

> Communism and dictatorship have been able to take over in countries only to the extent that people refused to accept and apply the teachings of God in their lives, and where they failed to take an active interest in public welfare. We must realize that the service we give is the rent we pay for the privilege of living in this old world, and the higher rent we pay, the better place we shall have to live in.[31]

The evangelical enthusiasm aside, Tanner's veiled message was that the province intended to bide its time. Although he publicly announced the government's oil sands policy would encourage immediate development, the province did not intend to bring the oil sands on stream for the foreseeable future. Tanner's decision to emphasize the role of the oil sands in North American security arrangements made the continuing development of the oil sands palatable to conventional producers by suggesting that synthetic production would only be tied to world events. If North America lost access to cheap offshore imports, the oil sands could serve as an effective backstop.

In any case, the terms of the proposed lease and royalty arrangements did little to encourage development. Since any firm looking to develop the oil sands must show tangible results within a year of taking out a lease, the short term risks would be large. Given the long lead-in time required to bring a commercial oil sands plant on-stream, it would be 1955 before production could begin. However, as Clark pointed out, "no company was willing to take on such an obligation [for] 1954–55."[32] Although many companies were interested in the oil sands as a future source of production, corporate timetables envisioned the oil sands arriving online much later than the mid-1950s. The prospective costs discouraged other companies from launching the full-scale mining program required to tap the sands.

Nonetheless, the Blair Report and the subsequent symposium were successes. In the words of historian David Breen, "everyone understood that locked within the sands was one of the world's largest oil deposits, if only the problem of a cost-competitive technology could be solved."[33] Therefore, by design more than by accident, the provincial government encouraged perceptions that the oil sands were a strategically important natural resource held in reserve, effectively delaying development and exploitation.

THE IMPORTANCE OF THE U.S. MARKET

Just when the threat from the oil sands began to recede for conventional producers, U.S. oil policy administered a staggering blow to the Canadian industry with one hand, but offered plenty of indirect support with the other. America's independent producers, concerned about the effect of foreign imports on U.S. crude prices, prevailed on the Eisenhower administration to impose mandatory import quotas.[34] Canada and Venezuela managed to gain an exclusion from the voluntary import quotas established by Eisenhower in 1955. Nevertheless, Canadian crude exports to the United States would remain static for the time being.[35]

But another Eisenhower policy encouraged American investment in the Canadian oil industry. In 1955, the U.S. treasury department decided that royalty payments made by multinational oil companies to host governments could be treated as payments of foreign taxes instead of business expenses for U.S. tax purposes. The change in U.S. tax law dramatically reduced the amount of tax paid by U.S. multinationals, and stimulated investment abroad. Historian Diane Kunz wrote that the royalty payments made by U.S. multinationals generally exceeded official U.S. foreign aid payments and enabled the Eisenhower administration to "privatize" its foreign and economic policies.[36]

Despite the valuable tax concession that encouraged investment in the Canadian industry, the voluntary quotas dashed hopes that the U.S. export market could absorb the influx of synthetic fuel. Furthermore, Canada's independent producers continued to regard the potential of the oil sands as a threat to their status and continued financial well-being. They continued to lobby the province hard to make sure synthetics did not usurp their position. Their concerns became particularly important after a restructured Great Canadian Oil Sands initiated discussions with Edmonton in 1954.[37]

For the newly reconstituted company,[38] the chief obstacle blocking development was Alberta's portioning formula. While the regulation of production from conventional wells was straightforward, the same was not true for oil sands production. Since the oil sands would be, largely, a mining operation, they would require large levels of capital investment and infrastructure to sustain their development. The rub for Great Canadian Oil Sands was that the large capital expenditures required to sustain synthetic fuel development might prevent oil sands producers from responding to market conditions. With higher fixed costs to cover, and economies of scale, synthetic producers could not cut back production in response to market conditions like conventional oil producers.

For conventional producers, Great Canadian Oil Sands' presentation to the Conservation Board confirmed the threat of the oil sands to their operations. The size and scope of the synthetic industry meant that the oil sands were like a huge grandfather clock. Once set in motion, business and government could only watch the clock unwind. Producers had to maintain production levels, despite market demand. Contrary to Clark's previous argument – that increasing Canadian reserves would stimulate growth across the petroleum industry – many inside the conventional industry believed that commercial development of the oil sands would have the opposite effect. Increasing production would simply depress world prices and reduce the profits of conventional producers already struggling to sell their product on the market.

Meanwhile, policymakers and company executives had to weigh the costs and benefits of bringing an expensive alternative source on stream at a time when Alberta's market was saturated. By 1955, Alberta wells operated with an average shut-in capacity of 30 per cent, meaning that 30 per cent of the oil produced by a given well would remain in the ground. Less than five years later, in 1959, Alberta's shut-in capacity would top 50 per cent.[39] Many independent producers argued strenuously that, to grow their companies, they needed to expand their market share. The independents already felt pinched financially when denied immediate return on past investments. Introducing the oil sands into the mix would aggravate the situation.[40]

Rather than allowing the issue to fester, the province moved to show its support for the independent producers, albeit in a backhanded manner. To the conventional industry's surprise, in March 1955, Premier Manning announced that production from the oil sands would be exempt from the pro-rationing mechanism administered by the Conservation Board. Furthermore, Manning dictated that any pipeline operating under provincial jurisdiction could not discriminate between oil from the oil sands and conventional oil, guaranteeing the oil sands a method of transportation to refineries.

In exchange, however, Manning also informed Great Canadian that since the portioning scheme excluded oil sands production, the province would *exclude* oil sands products from Alberta's existing markets. In other words, the province treated production from the oil sands as a commodity distinct from Alberta's conventional oil and gas production. As an independent agent, any oil sands company would have to find its own market because the province refused to include oil sands production as part of its informal quota. In effect, the provincial government gave synthetic producers the green light to begin production but ensured no operation could break even, let alone turn a profit. Members of the Legislature quickly approved the new policy when they passed the Bituminous Sands Act a month later.[41]

International developments temporarily gave Canadian crude increased importance. When war erupted in the Middle East in October 1956, the previously implied threat to European oil supplies became a reality. As historian Gerald Nash writes, "oil from any source came to be in great demand."[42] The closure of the Suez Canal prevented delivery of more than 1.4 million barrels per day of crude oil. Disruption to the Iraqi pipeline removed an additional 544,000 barrels per day from the market. Taken together, these events reduced world oil supplies by approximately two million barrels per day and eliminated whatever slack the system may have had.

Canadian oil already reached most of the markets readily available to it in North America but the disruption of the international market created an additional short-term demand. Markets further south, specifically those in California and along the Pacific coast – long the elusive goal of Canada's independent producers – suddenly became available. Tanker shortages along the West Coast made Canadian crude, delivered by pipeline, more attractive. However, at the conclusion of the crisis, the sales of Canadian producers dropped radically, and the world price of crude oil dipped.[43]

The pro-rationing system established in 1950 functioned well for seven years, but as market demand began to wane in early 1957, producers demanded that the province review the system. Although Alberta possessed enough capacity to supply all of Canada's petroleum needs, the province still sold less than half its available production. The contraction of the U.S. market at the end of the Suez Crisis exacerbated the situation. In December 1957, the American government announced another round of "voluntary" import quotas, adding insult to injury. This time, however, the reduction would apply to District Number 5 – the Pacific Northwest – previously exempt from import quotas and Alberta's largest market. Although the quota was set at 220,000 barrels per day – a level well above the oil exported by the province – the decision sent shockwaves throughout the beleaguered Canadian industry.[44]

Clearly, the main problem in the province remained the lack of available markets for Alberta's oil. Some observers thought the time was ripe to re-evaluate the continental orientation of the Canadian oil industry. This was the position of Premier Manning and independent producers, like Canadian Oil Companies, Dome Petroleum, Home Oil, Husky, Triad and Western Decalta. They began lobbying the federal government to open Montreal's extensive refineries to Alberta's crude and insisted that this had now "become a national necessity."[45] The task of re-evaluating Canada's energy strategy would fall to new Prime Minister John Diefenbaker.

THE BORDEN COMMISSION

On his second day in office, Diefenbaker announced that he would create a royal commission to investigate and make recommendations about the country's energy industry. Popularly known as "The Borden Commission," the inquiry took its name from the chair, Henry Borden, a Toronto industrialist and long-time member of the Conservative Party. The commissioners enjoyed broad authority to explore and make recommendations on a variety of subjects related to energy. Most notably, the Borden

Commission assessed and recommended "the policies which will best serve the national interest in relation to the export of energy and sources of energy from Canada."[46] Among those considerations included the possibility of building a pipeline from Alberta to Montreal, a market that would provide an outlet for another 200,000 barrels per day of Alberta crude.[47]

The commission conducted hearings on a variety of energy related issues beginning in Calgary on February 1958 and wrapping up in Montreal in July. In welcoming the panel to Calgary, Premier Manning reminded commission members that the province's energy industry had grown substantially in the previous decade. Yet, in spite of these gains, the industry had not reached its full potential. It was clear the Premier believed the commission could alter the *status quo* and sought to influence the direction the final report would take. Manning quickly outlined the basic problem facing Alberta's industry. The lack of available markets froze millions of dollars of investment capital "spent in drilling wells which are now capped." Without access to other markets, there would be little reason for companies to continue investing in the Canadian oil patch. The Premier asked the commissioners to provide "a clearly defined national policy with respect to both domestic and export markets."[48]

The province's excess capacity and the oil sands provided the means for Canada to become energy self-sufficient if the federal government was so inclined. However, carrying out a national energy strategy would have other costs. Building infrastructure – like pipelines, transportation, refining, and upgrading facilities – required significant capital that the federal government was unwilling to commit. During the acrimonious TransCanada Pipeline debate, the St. Laurent government revealed that the federal government would have to make an eighty-million-dollar loan to pay the construction costs of building the pipeline from Alberta to Manitoba. While the move would, in effect, subsidize Alberta's oil industry, it also suggested that Eastern consumers would have to pay higher prices for fuel. This view was reinforced by Montreal refiners, representing over one-third of the country's refining capacity, who believed that Canadian crude could not arrive in Montreal "by normal commercial means at prices competitive with those of foreign crudes."[49]

The fear amongst refiners was that if they were forced to bring in more expensive Alberta crude, they would be unable to compete with American refiners who sold cheaper foreign crude. Imperial Oil, the dominant firm in Canadian oil and gas, concurred with the refiners' position, arguing that the Montreal market was not the answer for Alberta's woes.[50] Implementing long-term throughput guarantees would ensure financing for the $200 million pipeline. The refiners would not provide these guarantees unless the federal government provided them with protection from cheaper foreign crude. Instead, the proper course would be to let the market determine policy. Given enough time, the "natural market" for Canadian crude, the United States, would open. Conventional producers countered with the claim that a healthy, vibrant oil industry was in "the national interest."[51] If conventional producers continued to shut-in their productive capacity – by now, nearly 50 per cent of all wells drilled in Alberta – many feared that the Canadian oil industry simply would not survive.

Although the oil sands were not the focus of the commission's enquiry, the Research Council made a presentation that summarized the enormous potential of the oil sands. The Council repeated Blair's conclusions and brought them up-to-date. It observed that, generally, operating costs and the value of oil products "appear to show similar percentage increases. In other words, the profit per barrel would still be quite similar to that shown in the Blair report."[52] While capital costs had increased, the Council pointed out that technological advances had also taken place to offset these hikes.

The submission did not entirely focus on the fiscal bottom line. Indeed, the Research Council suggested that much more was at stake than just the profitability of synthetic fuels. Rather, the future of Canada's oil sands hung in the balance. The presentation emphasized that it would take roughly three years after deciding to continue with commercial development before an oil sand project could yield oil. "The first successful [commercial plant] will be the result of sound judgment and correct timing," concluded the Research Council's authors. The conclusion left little doubt that it was the Province of Alberta, not private business, which would determine when the oil sands would come on stream.[53]

As the Commission held hearings across the country, Alberta received an unexpected proposal regarding the oil sands that helped to rekindle interest in synthetic fuel development. M.L Natland, a senior geologist with Richfield Oil Corporation, approached the Research Council of Alberta with a unique idea. While on a geological expedition in Saudi Arabia, Natland watched the sun set. He began to wonder if the energy from a small nuclear explosion could release the oil from the tar sands. "The major production problem is the natural viscosity of the oil which is hundreds of times greater than that of most other oils," surmised Natland. "It has long been recognized that the most promising way of freeing the oil from the sand would be to heat it." Following the argument to its logical conclusion, Natland wondered what the heat and energy created by a nuclear explosion could do to free the tar sands. "The tremendous heat and shock energy released by an underground nuclear explosion would be distributed so as to raise the temperature of a large quantity of oil and reduce its viscosity sufficiently to permit its recovery by conventional oil-field methods." Thus, in "Project Oilsand," a proposal formally submitted to the Research Council, Natland suggested that Richfield Oil Corporation detonate a nine-kiloton nuclear warhead underground to test his hypothesis (the yield of the atomic bomb dropped on Hiroshima was estimated at 12.5 kilotons).[54] If successful, Richfield's scientists believed they might have a way to "create an oilfield on demand."[55]

What is striking about Richfield's proposal is that it marked the beginning of a transition period in the development of the oil sands. For the first time, companies began seriously researching and investigating comparatively inexpensive *in situ* (literally, "in place") extraction methods. The first recorded experiment with *in situ* extraction in Alberta took place in 1929 when J.O. Absher began developing a very crude process. Absher drilled a borehole into the deposit, introduced a gasoline vapour/air mixture into the chamber and set it ablaze. A chromel-alumel thermocouple registered 2000° F before failing. The fire burned for two days before, in the words of Karl Clark, "things went wrong." The heat from the fire was so intense that pipes inserted into the hole became caked with paraffin and soot, rendering the well useless.[56]

Regardless of the failings of Absher's experiment, *in situ* processes remained attractive for several reasons: they did not require companies to remove layers of overburden and muskeg from the deposit before mining began. They also held the promise that they could deliver oil sand oil to the surface like conventional crude. Perhaps most important of all, with *in situ* production, developers were no longer limited to the 2 per cent of the Athabasca deposit accessible by conventional mining methods. The potential productive capacity of the oil sands would leap from 700 million barrels to 800 *billion* barrels.

Originally called "Project Cauldron," Natland's proposal received serious consideration from all levels of government in both Canada and the United States. On May 9, 1958, executives from Richfield Oil Corporation met with representatives from the United States Atomic Energy Commission to discuss the idea of using the nine-kiloton bomb as part of a test explosion to see if the idea was feasible. The USAEC approved the idea, and Richfield proceeded with the backing of the American government. With the first hurdle out of the way, representatives of the Richfield Corporation met with the Alberta government and began discussions with the federal government in June 1958. With the active support of Ernest Manning, Richfield continued pursuing the plan. After initial tests in Nevada were successful in early 1959, development of the project continued until July 1959 when the Joint Committee finally shelved the plan in deference to Canadian concerns about the use of nuclear energy.

Had the Richfield experiment proceeded and successfully liberated the oil from the oil sands, one has to wonder how Alberta's conventional producers would have reacted. The province still did not have a market for much of its existing conventional crude. To make matters worse, another large producing field in the Swan Hills region north of Edmonton came on-stream in December 1958. The province, literally awash in oil it could not sell, confronted the possibility that even more oil could glut the world market. The consequences were exciting and frightening at the same time. If successful, the Richfield test could, "at a single stroke double the *world's* petroleum reserves."[57]

In the meantime, the Borden Commission still had not issued its findings. Lacking a final report, the federal government gave no indication about which direction the nation's oil policy would take. For six tense months, all of the interested parties simply bided their time and waited for Ottawa's signal. The Borden Commission's final report proposed that very few changes take place in the Canadian industry. In part, the Commission based its findings on the assumption that Canadian crude was at some competitive disadvantage *vis-à-vis* other world sources. The cost of exploration, development and production of crude oil in Canada was higher than in Venezuela or the Middle East. In this regard, the Canadian production costs bore a close resemblance to those in the United States. "The combination of these circumstances puts Canadian crude, in effect, at a disadvantage in world markets and *limits possible export markets to the United States*," something Borden considered Alberta's "natural market."[58]

Later in the report, the commissioners discussed the possibility of developing a continental energy strategy. "Canadian and United States crudes would be freely used in refinery areas on the North American continent, supplemented by such imports of foreign crude as might be necessary to augment any shortage of supply from North American sources." The report made it clear that a continental energy policy might take a long time to emerge. However, the report attempted to preserve the opportunity that such a policy might occur. Canada's national oil policy, declared the commission, should be to encourage the growth and development of the Canadian oil industry's continental orientation. The industry, urged the report, should be encouraged to "take vigorous and imaginative action very substantially to enlarge its markets in the United States."[59]

Some historians later hailed the *Second Report* as a master of political compromise that satisfied all interested parties.[60] The reality is that very few segments of the industry demonstrated any real enthusiasm for the report. Oil majors thought the report reasonable; the independents remained unconvinced that the growth potential estimated for the Ontario and American markets were realistic, and resigned themselves to an ignominious end. The government of Alberta viewed the report with disappointment. Very few elected officials in Edmonton expected their federal counterparts to alienate voters in Ontario and Quebec for the sake of Alberta oil.[61] However, when faced with a choice about the industry's future, the federal government decided to place its faith in the American market. The development of the Canadian oil industry along continental lines continued unabated.

Several separate, but nonetheless related, forces aided the closer integration of Canadian and American markets. The heated nature of the debate in Canada over the construction of a pipeline from Alberta to Montreal prompted a reassessment of the import quota system in the United States. The reconsideration resulted in Canada's exemption from a redrafted system in 1959. This was exactly what Alberta's provincial government hoped would happen and they continued to press for the pipeline as leverage against the United States. The argument was stunningly simple: if Canadian oil remained subject to restrictions, then Canada would be forced to adopt a national strategy that would curtail Venezuelan crude imports to the Montreal market.[62]

The highest levels of the Eisenhower administration undertook an evaluation of the Canadian oil industry in late 1958. They concluded that the Canadian oil industry clearly could not compete with cheaper sources of foreign oil. The National Security Council acknowledged that the current limit on Canadian imports was not "high enough to stimulate exploration and development of Canadian resources." Rather than allowing the industry to stagnate, the NSC outlined a course of action that would justify removing import quotas on Canadian oil.[63] Figuring prominently in the NSC's argument was the potential threat of a transcontinental pipeline that would give Alberta oil access to the Montreal market at the expense of Venezuelan crude. Such a move would have deleterious effects on the Venezuelan petroleum industry. The Venezuelans would then blame the loss of the Montreal market on the intransigence of the United States.[64]

Excluding Canada from the mandatory import quota system did what the Borden Commission could not: it returned confidence to the Canadian industry. In a single stroke, the exemption stimulated investment and breathed new life into the oil patch. The Canadian industry also received another shot in the arm when the Texas Railroad Commission – the state agency that exercised control over the production of oil from the vast Texas fields – began reducing the level of allowable production from Texas wells. For Sun Oil, which had a major stake in the oilfields of West Texas, the decision prompted a search for foreign crude sources.[65] J. Howard Pew, the

President of Sun Oil, fervently believed the oil sands would become the company's most important fuel source, and doggedly pursued commercial development of the deposits. When commercial development of the oil sands began in the early-to mid-1960s, it was the result of a fortuitous combination of American interest, American capital, and the access to American markets.

CHAPTER 5

"Within Reach" and "Beyond Reach" Markets: The Reluctant Expansion of the Oil Sands, 1960–69

No nation can long be secure in this atomic age unless it be amply supplied with petroleum. It is the considered opinion of our group that if the North American continent is to produce the oil to meet its requirements in the years ahead, oil from the Athabascan area must of necessity play an important role. – *J. Howard Pew, President, Sun Oil*[1]

The start of commercial production of synthetic crude from the Athabasca tar sands has been hailed as the dawn of a new era, the forerunner of vast new supplies of hydrocarbon energy, assurance of hemispheric self-sufficiency in petroleum, and a threat to conventional crude oil.

It may turn out to be all of those things, but only time will tell. – *Oil and Gas Journal*[2]

As the 1960s began, Alberta's approach to oil sands development was still much the same as it had been since the early 1950s. The oil sands were an expensive alternative fuel operating on the margins of the oil industry when compared with the cheaper costs of producing conventional fuel. The province continued to worry that development of the oil sands could harm the interests of conventional producers. The dilemma facing oil sands producers remained the same for much of the 1960s. Without increased markets, further commercial development of oil sands would be impossible.

Ironically, this rigid and cautious approach to oil sands policy produced some long-term advantages. The provincial government was concerned about the deleterious effects that excessive production from the oil sands would cause the conventional industry. Edmonton set up long-range policies that would secure Alberta's supplies well into the new millennium. Caution thus afforded politicians and policymakers the necessary time and space to make important decisions. Instead of relying on the oil sands to produce large amounts of oil to meet a particular emergency, Great Canadian Oil Sands' development continued at a leisurely pace. This methodical approach was the result of an unspoken compromise between Sun Oil and the Province of Alberta. It also suited the needs of officials in Edmonton who developed a flexible oil sand policy with long-term benefits and few short-term dislocations.

J. HOWARD PEW AND SUN OIL

Sun Oil's interest in the oil sands actually began during the Second World War. A planning group in Sun's Philadelphia office investigated the patterns of future hydrocarbon consumption and production in the United States. The line of decreasing production intersected the line of increasing consumption at the mid-1960s. At that point, the group identified three alternative sources for the company to pursue to replace the decline in conventional production: oil shale, bituminous sands, and coal liquefaction. After consulting with Sun's upper management, the company decided to meet the demand through the oil sands. Consultations began with the Geological Survey of Canada in the spring of 1944 to gather information.[3]

The project remained on the backburner until George Dunlap became manager of Sun's Canadian operations in 1949. Before taking on his new assignment in Calgary, Dunlap travelled to the corporate head office to receive last-minute briefings and instructions. He had an unusual meeting with Sun's chairman, J. Howard Pew. "I understand you are going to

Canada," said Pew, and when Dunlap replied positively, Pew began to outline his personal views on Canadian petroleum reserves. The potential for Canada's natural resources was great, but J. Howard confessed, "I have one area that I'm interested in and would like to share with you my interest." Pew went to a cabinet and pulled out a thick file marked "Athabasca Tar Sands." File in hand, Pew revealed his deep personal interest in the sands.

The chairman told Dunlap that he believed the sands would, some day, be very important for Canada and North America. Pew confessed that he did not know what other members of Sun Oil had told Dunlap, but for him, Dunlap only had one important mission: to ensure that "Sun Oil Company always has a 'significant position' in the Athabasca Tar Sands area!" Dunlap replied that he would do his best to carry out Pew's instructions.[4] Perhaps still dazed by the force of Pew's convictions after the meeting, Dunlap relayed the thrust of his conversation to other members of Sun's executive team. Did they know what Pew had asked him to do? They replied that they were well aware of J. Howard's interest in the oil sands and advised Dunlap to "follow [Howard's] instructions and keep us fully informed about Athabasca."[5]

Dunlap's move to Calgary coincided with the province's initial efforts to interest the petroleum industry in the oil sands.[6] Shortly after the Blair Report's release and the province's Oil Sands Conference, Edward (Ned) Gilbert, Sun's Division Land Manager in Canada, wrote to Dunlap. The federal government's core drilling program identified 200 million barrels of recoverable bitumen *per section* in a four-section block in the Athabasca region. Gilbert strongly urged Dunlap to take out a lease in the area. For Sun's landman, the decision was as simple as it was straightforward. No other location in the world existed where the company could get "proven reserves for less than the $1.40 per acre they would cost in Fort McMurray."[7]

On Gilbert's recommendation, Sun took out a lease around the Bitumount location. Despite the initial burst of enthusiasm, early field data suggested that their lease would not be as productive as it might be elsewhere in the deposit. Not wanting to abandon the sands altogether, Gilbert suggested that the company approach Abasand Oil and try to take the more promising Lease #4 off their hands.[8]

From Abasand's point of view, Gilbert's suggestion could not come at a better time. Strapped for cash, Abasand had farmed out part of the lease to Canadian Oil Company but retained a portion for itself.[9] In exchange for Sun's capital, and technical expertise, Abasand traded their 75 per cent interest in their lease at Mildred-Ruth Lakes. The deal was good for both companies. It enabled Abasand to fulfill the terms of its lease with the provincial government by continuing development, while granting Sun access to the most valuable deposit in the McMurray formation.

For the next eight years, Sun carried out core drilling, continued research on heavy oils, and generally followed the technical developments affecting the development of the tar sands. In total, Sun invested just under half a million dollars in the oil sands, giving them a stake in the industry's future.[10] Sun continued to expand its interest in the oil sands throughout the 1950s. In 1958, it entered an agreement with Great Canadian Oil Sands. By the terms of the agreement, GCOS would use its technology to develop an integrated operation on the Ruth Lake lease at Tar Island. Sun would purchase 75 per cent of the output at a contracted price.[11]

By the early 1960s, several factors helped convince the province to give renewed emphasis to oil sands development. Despite massive additions to Canada's conventional reserves, the reserves-to-production ratio had fallen from forty-two to nineteen years of supply.[12] The problem of finding markets for Canadian crude that occupied much of the time and attention of the Borden Commission gradually waned by the end of the 1950s. A series of piecemeal market increases, mostly in domestic sales, absorbed more production. These increases were due, in no small part, to the fact that Canadians consumed more oil than ever before. In 1947, Canada's daily requirement of crude oil from all sources totalled 267,000 barrels. That total rose to 742,000 barrels per day in 1957. Over that same period, Canadian producers went from supplying less than 10 per cent of Canada's daily needs to providing approximately 47 per cent. With investments totalling over $3 billion in ten years, the conventional industry now assumed an important role in the economies of both the province and the country at large.[13]

In early January 1961, Canadian petroleum exports to the United States increased 71 per cent, or 149,482 barrels per day, over the first quarter of 1959. Canadian diplomats continued to lobby Washington to increase their imports of Canadian oil. The marketing problems that hampered previous attempts to begin commercial development of the oil sands seemed outdated.[14] While the province would never totally abandon its conventional producers, oil sands policy was clearly on the verge of shifting directions. Keeping to its earlier pledges of support for the conventional industry, the province continued to emphasize the marketing of conventional oil even as it encouraged oil sands development.

At the highest levels of the provincial government there was recognition that the two objectives might conflict. Premier Manning was savvy enough to appreciate that the oil industry was the province's golden goose. As such, the provincial government made it clear that commercial development of oil sands, if it occurred at all, would proceed under guidelines established by the provincial government. Alberta would not permit synthetic production to displace large amounts of conventional production.[15] Part of Manning's success in pursuing these conflicting policies stem from the particular dynamics and pressures inherent in Alberta's provincial politics. Many observers have noted that the Alberta electorate is unusually forgiving and allows political parties to rule the province for decades on end. Since 1935, Alberta's government has remained conservative and staunchly pro-business.[16] "Albertans," wrote political scientist Brooke Jeffrey, are "accustomed to one-party rule and lengthy dynasties."[17]

Perhaps the tendency to grant provincial politicians unusually long terms in office offered Alberta's leaders a luxury not always afforded elected officials: time to plan and implement long-range strategies. The province's successive Social Credit governments had time to balance short-term goals against long-term objectives. Thus, even if the province was dependent on conventional oil and gas revenues, Manning pushed forward with oil sands development in the 1960s for two reasons. First, the Premier remained personally committed to the development of the oil sands and securing the province's economic future. Second, his electoral popularity meant the premier could afford to take the risk. Manning did not need to worry that voters would turn his government out of office if it failed in its attempts to develop the first commercial oil sands plant. In fact, the only real risk was that the U.S. market would accept more Canadian oil when the first commercial plant began to produce large amounts of fuel. To reduce this risk while pushing forward, the province favoured construction of a small-scale commercial plant. By operating a small-scale plant, he believed that the company could rapidly step up production should the need arise. However, according to the premier, to go beyond one plant at this stage would "only aggravate a marketing situation which is already serious."[18]

The only company positioned to take advantage of the government's new policy was Great Canadian Oil Sands Limited (GCOS). Compared with other firms looking to begin commercial development of the oil sands, GCOS's plans were conservative. In its first application to the Conservation Board, GCOS proposed to build a small plant capable of producing 11.5 million barrels of synthetic crude annually (31,500 barrels per day.) Using a variation of Clark's hot water separation method, they would mine the sand via a bucket-wheel excavator and transport the material by conveyor system to the extraction plant. After the primary separation took place, a pipeline would transport the oil to Edmonton. Great Canadian Oil Sands claimed it already possessed a market for its product: Sun Oil agreed to purchase approximately 24,000 barrels of the plant's projected daily total. Sun would then ship the synthetic crude to its refinery in Sarnia, Ontario for processing and eventual distribution.[19]

As hearings on the proposal began in June 1960, Great Canadian Oil Sands would clearly have a difficult time meeting the Conservation Board's rigorous criteria for approval. The Board acknowledged the desirability of proving the viability of the oil sands as an alternative fuel source. However, it weighed those considerations against the impact synthetic crude would have on the conventional market. The Board would not limit the scope of its inquiry in any way. Oil sand proposals were therefore evaluated according to three separate criteria: conservation, the impact on the conventional market, and the economic and technical feasibility of the operation.[20] The Conservation Board understood that it was on shaky ground on this final point. No real way existed to prove the project's economic or technical feasibility. Commissioners therefore gathered and assessed as much data as possible before rendering a decision. The Board understood that if it approved any commercial oil sands plan that ultimately failed, "serious discredit might result which, in turn, might delay their development for many years."[21]

Significantly, the Conservation Board concluded, "synthetic crude oil supplied from the Great Canadian plant would cause a reduction in demand for conventionally produced crude oil." Most of the displaced oil would be from the Province of Alberta. "Approval of the Great Canadian application would cause a loss of approximately 5 per cent of the total demand for conventionally produced crude oil in Alberta, or some twenty to 30 per cent of the pro-rateable market demand," concluded the Board. Great Canadian and Sun Oil did produce small amounts of conventional crude in the province, but their share of Alberta's pro-rateable market was negligible. Other producers would bear the net effect of introducing production from the oil sands, especially those with wells "in the more prolific pools."[22]

Ultimately, the Conservation Board fully realized that production from the oil sands would be desirable "some day." But the present proposal from Great Canadian Oil Sands was currently not in the best interests of the province. Another proposal from Great Canadian Oil Sands, however, would be welcome before June 30, 1962. In closing their remarks, the Board suggested that the new plan should show "a less significant effect on the market for conventionally produced Alberta crude oil."[23]

The delay enabled the provincial government time to develop a strategy for orderly development of the oil sands. It also recognized the delay would buy time for the province's struggling conventional producers to solidify their markets and develop strategies to cope with the competition provided by the oil sands. With more than 40 per cent of Canadian demand being met through imports, and producers in the United States demanding further restrictions on Canadian exports, Alberta's conventional producers faced a grim future. Spurred along by the impending GCOS hearings, eighteen of the smaller independents joined together to form the Independent Canadian Petroleum Producers (ICPP) – the precursor to the Independent Petroleum Association of Canada – to lobby Edmonton and Ottawa for a more proactive oil policy.[24]

The decline in revenue from oil and gas royalties illustrates the challenge facing the provincial government and the conventional industry in the early 1960s. From a peak of more than $37 million in 1957, Alberta's conventional oil and gas royalties fell approximately $12 million to $25 million in 1959. Because of these dire circumstances, and the better organized response by independent producers, the province's interest in developing the sands waned appreciably. In late February, 1962, just weeks before GCOS presented its amended plant to the Conservation Board, Premier Manning warned that the province might delay development of the sands indefinitely. Perhaps, mused the premier, the province would delay synthetic fuel development until stability returned to the world market. The determining factor would be the search for suitable markets that would not harm conventional producers. "There must be a daily guarantee for production," reasoned Manning, "and not the fluctuations which exist today." Further undercutting the position of oil sands producers was the recognition that synthetic crude did not offer consumers either a price or a quality advantage over conventional crude. In fact, it was possible to suggest that the price of conventional crude was much cheaper. Perhaps, the province could find a market for the oil sands by 1975.[25]

Opposition critic Mike Maccagno blasted the government's stance in the Legislature, perhaps believing that Manning had telegraphed the government's position on GCOS's revised application to the Conservation Board. Although the premier's concern for market realities had some merit, the province could not afford to wait indefinitely, countered Maccagno. If the development of the oil sands did not go forward, research and development of alternative fuel sources – most notably nuclear energy – would render the oil sands useless. "Surely with the present demands increasing each year, it ought to be possible to secure a small part of the available market, say just five per cent, just to prove the development of the sands can be done."[26]

In March 1962, GCOS submitted an amended application to the Conservation Board. Beyond the political turmoil swirling in the province, Great Canadian also had to contend with the fact that it was no longer the sole applicant. Two other companies, Cities Service and Shell, had also submitted applications to the Conservation Board. In evaluating GCOS's amended proposal, the Board established three useful criteria to assess the impact of oil sands production on the conventional industry: the ratio of actual production to productive capacity, the conventional industry's life index (the number of years current production would support levels of current consumption), and the pro-ration allocation factor (the fraction of productive capacity over the economic allowance they permit a well or pool to produce besides its allowance.)[27]

The Board did express some reservations about GCOS's ability to operate profitably given the proposed plant's maximum daily output of 31,500 barrels per day. Nevertheless, the Board gave Great Canadian a chance to see if the low production total was economically feasible.[28] After much consideration, the Conservation Board recommended approval of GCOS's application and the government concurred on October 2, 1962. Finally, after nearly five decades of research, the province was going to begin commercial production of the oil sands with the help of a foreign-owned, integrated oil company.

It is not difficult to understand the province's desire to see production from the oil sands enter the market. It is more difficult to explain Sun Oil's involvement as part of a sound business strategy. In fact, sound business practices might have argued against the company's involvement with the oil sands. The technology associated with the Great Canadian proposal was unproven on a commercial scale. Furthermore, synthetic crude production meant high fixed costs for production. With no readily identifiable market for oil sands production, the likelihood of Sun recouping its initial investment in the near future seemed remote. Clearly, something other than a concern for the corporate bottom line was the determining factor in Sun's involvement with the oil sands.

As the head of both Sun Oil and the Pew family, J. Howard Pew strongly believed that the oil sands represented a unique challenge. This was the kind of opportunity the company should willingly accept. Sun Oil had long charted an independent course in an industry known for creative wildcatters.[29] From the earliest days of the twentieth century, Sun forged a reputation for entrepreneurial spirit and drive. J. Howard Pew inspired the company's trailblazing refining techniques, enabling the company to convert poor quality, heavy East Texas crude into high-quality lubricating oil. The construction of the "Big Inch" pipeline in the 1930s helped further to cement Sun's reputation as industry pioneers. As one company historian has written, J. Howard Pew displayed an uncanny knack for anticipating changes in the business environment and positioning his company to meet them. By the mid-to late-1950s, the change in the industry appeared to Pew to point directly to synthetic fuels.

Perhaps, however, Pew's enthusiasm for the oil sands reflected more than mere corporate adaptation to changing circumstances. The oil sands were a challenge for Sun to overcome. "Unless projects of this kind were periodically challenged and solved," reasoned Pew, "our organization would become soft and eventually useless."[30]

Beyond the technical challenge of oil sands production, Pew's solid friendship with Premier Ernest Manning nurtured his interest. Both Manning and Pew shared a similar private enterprise philosophy. They believed that business leaders and government officials had a duty to exhibit self-discipline and social responsibility. Pew described the responsibilities of management as a form of "trusteeship," where management was accountable to stockholders, employees and to the public at large.[31] Similar attitudes toward the roles of business and government extended to their private beliefs. Manning was a Baptist lay preacher whose religion so deeply informed his political views that commentators cannot divorce the two. Pew told the Presbyterian Men's Council that Manning preached "one of the greatest sermons I have ever heard." Even the two men's wives shared a special bond, generally spending summers together in Jasper National Park.[32]

THE OIL SANDS DEVELOPMENT STRATEGY

Perhaps the most accurate indicator of the Pew/Manning friendship came in the aftermath of the Conservation Board's approval, when the province outlined its new oil sands strategy. Manning cited an obvious responsibility to regulate the timing and extent of oil sands production to protect the interests of both the public and the conventional oil industry. Nevertheless, the Manning government made it clear that it intended to continue carefully with development to ensure that oil sand production would "supplement but not displace conventional oil." The conventional industry provided a good bounty for the province. Now the province would act to preserve "the stability of conventional oil development" and provide "the necessary incentive to ensure its continued growth." Manning repeated the decision that oil sands production could continue unabated, provided it could reach markets "clearly beyond present or foreseeable reach of Alberta's conventional industry."

Meanwhile, the province would restrict oil sands production to 5 per cent of the total demand for Alberta's oil – approximately the same output approved by the Conservation Board for GCOS. While there would be some flexibility in instances when the world market made changes necessary, Manning made it clear that the province's primary responsibility was to keep conventional producers happy.[33]

Manning's policy declaration jeopardized the applications of Cities Service and Shell. With the province implementing a 5 per cent ceiling on oil sands production, little room existed underneath the cap for other projects. This was particularly true for projects that proposed to increase the province's daily production by 100,000 barrels per day. Pegging the province's total allowable oil sands production so closely to the target established by Great Canadian guaranteed GCOS a monopoly. Vernon Taylor of Imperial Oil, part of the Cities' Service group, awaiting a ruling from the Conservation Board, chose his words carefully after Manning announced the policy. Taylor thought the policy reasonable but hoped the province would increase the market share allocated to the oil sands "depending on how the economics of oil sands production develop."[34] Meanwhile, influential *Oilweek* publisher C.V. Myers offered support for the policy, mostly because of the protection it afforded conventional producers.[35]

Despite Edmonton's clear signals, Cities Service and Shell pressed forward with their applications. The two companies hoped the Board would be willing to interpret government policy more liberally and expand the oil sands' share of the market accordingly. But the prized concessions were not forthcoming. The Conservation Board considered what the total impact of synthetic crude production would be on the conventional industry. In its calculations, the Board assumed that all synthetic production, including Great Canadian's 31,500 barrels per day, would displace supply from the province's conventional industry and would not capture any "new" markets.

Since both the Cities and Shell proposals sought approval for facilities designed to produce approximately 100,000 barrels per day, the Board found itself in a tenuous position. If the Conservation Board approved all three projects, oil sands production would represent between 18 and 19 per cent of the province's daily output. The numbers hardly improved when the possibility of approving one of the two projects was raised. If the Board approved either the Cities Service or Shell application, the conventional industry would be unable to recover its lost market share until 1973 at the earliest. Clearly, the percentage of the market allocated to the oil sands if the Board granted either of the applications was "well beyond that contemplated under the first clause of Government policy."[36] George Govier wasted few words in turning down the two applications. The policy clearly stated that the conventional industry should not suffer because the province embarked on a synthetic fuels program. "To withhold the growth opportunity until 1973, 1975 or later would in the Board's opinion be contrary to the policy."[37]

The most important question that remained was how much latitude the Board had to enforce the province's limit on oil sands production to 5 per cent of all crude production. Could producers expect the Conservation Board to free some additional room in the province's production quotas for them to operate? If not, when could they reasonably expect the province to approve their applications? On this issue, the province mitigated the Board's discretion, determined as it was to minimize the conventional industry's market losses. Given this constraint, it was apparent the Board could not even approve another small project like Great Canadian, let alone the larger applications before it. Nonetheless, the Board believed that the province's new policy favoured growth both in the conventional and synthetic industries.

Lacking further instructions from Manning or his ministers, the Conservation Board interpreted the government's policy quite rigidly. After assessing the supply and demand capacities of Alberta crude, the Board estimated that the most reasonable solution lay in dividing "excess market growth" equally. This way, the conventional industry did not lose the share of markets it already possessed and the government could still legitimately claim to be encouraging synthetic fuel production. The province continued to favour conventional production by guaranteeing a growth rate of no less than one percentage point per year for conventional producers. The Board then presented the applicants with a forecast of total "permissible" levels of production for the oil sands. Any proposed plant with a capacity of 100,000 barrels per day would "not be consistent" with the Board's interpretation of government policy "until about the end of 1976."[38] Although both applicants were encouraged to resubmit their applications in 1968, Great Canadian clearly remained the chosen instrument of the government's oil sands policy.

Committed as it was to the development of the oil sands, the province took steps to reduce other financial uncertainties even as the GCOS application made its way through the lengthy application process. Although Manning hoped to clarify the royalty situation, the flexible royalty strategy his government adopted muddled matters even more.[39] The Mines and Minerals Act of 1962 set the Crown royalty on the production of synthetic crude at 1/6 rather than the rate of 1/10 assumed by Great Canadian in the application. GCOS would now have to pay royalties of 8 per cent on crude production up to 900,000 barrels and 20 per cent on any subsequent production. There was an additional 16.66 per cent royalty on production of refined and specialized by-products of the bitumen.[40]

The province ignored pleas to simplify its royalty regime. Premier Manning sympathized with GCOS, but suggested that the province would not alter its royalty structure at this time. Perhaps "three, or four, or five years" after production began the province would revisit the royalty issue, but for now Great Canadian would simply have to accept the rules. Manning also responded to critics who charged that production from the oil sands was too expensive and would not find a market. The premier agreed that the cost factor still made development of the oil sands a "borderline case" but reaffirmed Edmonton's commitment to the project. Development of the oil sands would continue because several questions remained about the viability of the sands as an energy source.[41]

Having seen the government's position, it now fell to the industry to see if the system would work. In part because of the new royalty system, the three primary investors in GCOS – Canadian Oil, Ltd., Canadian Pacific Oil and Gas, and Sun Oil – contracted Canadian Bechtel to prepare an independent review of their project. The need for such an assessment was readily apparent. By the end of the application procedure, the financing provisions outlined in the application were hopelessly out of date. Circumstances had radically changed. GCOS had to account for the new royalty provisions outlined by the province and Edmonton expected advance royalty payments for the first eight million barrels. The municipality of Fort McMurray also sought to benefit from the fledgling industry, levying an annual municipal tax of $500,000 on GCOS. The final indignity came a few months later. The federal government removed the company's exemption on an 11 per cent sales tax for all building materials, equipment, and machinery purchased or imported into the country.[42]

Beyond the rapidly changing economic picture, there were other important reasons to undertake feasibility studies before beginning the project. Previously, three specialized contractors provided independent estimates on different aspects of the Great Canadian project. While there was a general appreciation of how the whole project fit together, it was now time to begin the more concrete work of piecing the parts together. After consulting with several industry sources, the unanimous conclusion was that GCOS should hire Sidney M. Blair's firm, Canadian Bechtel, to assess the entire application rather than an individual section.

The engineering firm began the process of assembling and assessing information and the early results were less than encouraging. Interim reports as early as April 1963 suggested that there was little wrong with the plant's engineering. However, even the most preliminary economic data indicated that a plant with a capacity of 31,500 barrels per day was simply unfeasible. The only plant size that would give a reasonable rate of return to investors would be one with a daily capacity of 45,000 barrels per day. After consulting with the other partners, Canadian Bechtel prepared an estimate for plants of each size.

When Bechtel released the results of the review in May of the following year, they were not very promising for the 31,500-barrel-per-day plant. Given the constraints imposed on production by the Conservation Board, it would take more than ten years before the company would turn a profit. If, however, the partners were willing to invest in the larger plant, Bechtel believed the investors could expect to recoup their investment within eight years of project start-up. However, Bechtel warned that the increased plant size and capacity came with a hefty price tag. It would require an additional $96 million simply to build the larger plant. Frightened by the numbers, Canadian Pacific and Shell Canada – who had taken over Canadian Oil's share in the interim – believed the risks were too high and dropped out of the project. Sun, however, announced that it "was prepared to make every effort to complete the project."

Company officials told GCOS that it was willing to back up its words with $67.5 million, on two conditions. The first was that Sun Oil would increase its share of GCOS from 33 to 80 per cent. Secondly, Sun insisted that the project would only continue if the Conservation Board would expand the size of the project by 14,500 barrels per day.[43]

The increase in production would push Great Canadian's daily production to 45,000 barrels per day. The increase represented approximately 7.5 per cent of the province's crude production – just outside the 5 per cent parameter established by the province's guidelines in October 1962. If the province denied GCOS's request, Sun Oil made it abundantly clear that they were willing to remove their support from the project as well. Without Sun's financial support, Great Canadian would clearly collapse in a sea of red ink.

There was nowhere else for GCOS to turn. Since June 1, 1962, GCOS had issued 549,987 common shares to help finance its oil sands project. Sun Oil had purchased just under half those shares, giving the Philadelphia-based company an important voice in GCOS's future. As business historian Graham Taylor later concluded, "No Canadian company – and for that matter, no American company other than Sun – was prepared to take the risks inherent in this venture."[44] Having deferred judgment on the only other oil sands proposals until 1968 at the earliest, the Board faced the possibility that commercial production would not begin until early- to mid-1970. As GCOS amended its proposal for a second time, the future clearly hinged on how rigidly the Conservation Board interpreted the 5 per cent provision of the policy.

Great Canadian Oil Sands presented its amended proposal to the Oil and Gas Conservation Board in early September. In its brief, Great Canadian argued that the Board should approve the proposed increase for four reasons:

1. The Conservation Board had overestimated conventional production levels from other provinces.

2. The unfavourable balance of Canadian trade would encourage the federal government to introduce as much Canadian oil as possible to Eastern markets.

3. The U.S. balance of payments problems combined with its diminishing rate of growth in crude reserves would encourage

additional imports of Canadian crude.

4. GCOS had obtained a "new market" for synthetic crude and, as such, a good portion of the additional production was exempt from the province's pro-rationing formula.[45]

In his testimony, Sun's President stated that, pending approval by the federal governments of both the United States and Canada, Sun would be willing to take 10,125 barrels of the additional capacity to its Toledo refinery. This, argued Dunlop, classified as a new market since Sun had only run Canadian crude through it once before on a special occasion. Although Dunlop conceded that this was not a market "clearly beyond those supplied by conventional crude," he said that Sun Oil considered it one nonetheless.[46]

Not surprisingly, GCOS's application outraged Cities Service. In Cities' opinion, the terms of the amended proposal went well beyond that necessary to offset changes in the economic environment. Cities Service President A. P. Frame stated that the differences between the approved application and the present one were such that the latter became "a new application in itself."[47] GCOS had the opportunity to decide the size of its proposed plant and erred on the side of caution "to establish a status before the Board." Frame argued that the company should have to live with its mistake. Furthermore, the revised application established a dangerous precedent and, in Cities' opinion, made it plain for all to see that GCOS was not honest. If the Board granted GCOS's revised production level, Cities inferred darkly that it would not be the last. "A series of small bites is going to be easier to get than one large gulp," warned Frame.[48]

In making its decision, the Alberta Conservation Board considered each of Great Canadian's arguments, and dismissed them. The Board conceded that its estimates of Canadian production were high, but not so much that they should accept Great Canadian's revised figures. GCOS was on firmer ground with its argument about the Canadian balance of trade. However, the Board pointed out that this was a matter of federal government policy over which the Conservation Board exerted little to no influence. The Board dismissed the argument about the U.S. balance of payments problem on similar grounds. Future penetration of U.S. markets would depend on U.S.

government policy, not Conservation Board pro-rationing.[49] Finally, the Conservation Board had to consider Sun Oil's contentious claim that the Toledo market formed a "new market" for Canadian crude. "The boundaries to such markets are geographical," concluded the report, and were not "defined according to individual company policies within an existing market area." The Conservation Board held that Toledo was clearly within the present and foreseeable reach of Alberta's conventional industry and did not form a "new market."[50]

The Board then turned its attention to Cities Service's claim that the amended application constituted a new proposal and that the original GCOS proposal had been misleading. To determine the application's status (i.e., whether or not the amended GCOS proposal constituted a new project), the Board suggested that "one should look primarily at the project itself rather than the effects the proposed changes would have outside the project." From this point of view, the Board ruled that the basic character of the proposal remained intact in spite of the increased production levels. While the Board agreed that the question of scale was important "in the broader consideration of the application," taken together, the proposed amendments did not make up a new application.[51] The Board reserved some of its strongest criticism for Cities' second charge. Govier and his colleagues largely agreed with Cities Service's contention that Great Canadian's amended proposal refuted the original cost evidence presented in 1962. The Board expressed unease over GCOS's earlier figures and said it was "disturbed" that the company had based its proposal on unrealistic figures. Despite these reservations, the Board concluded that the original application was not deliberately misleading. It adopted the view that the inaccurate cost conclusions reflected unbounded optimism and incomplete study.[52]

Ultimately, the Conservation Board approved the amended GCOS proposal because it was the only one that did not propose a massive dislocation of conventional crude sources. *Oilweek* accurately reflected the opinion of the industry when it called the Board's decision to allow GCOS to increase its production "the best of a bad situation."[53] If GCOS dropped out of the picture, the only alternative available to the province would be to approach Cities Service or Shell to continue with their applications. Given the wide disparity in production proposed – 45,000 barrels per day versus

100,000 barrels per day – perhaps the Board's decision was easy. In any case, GCOS did not leave anything to chance and played this trump card as well at the hearings. One GCOS executive pointed out that the government announced the 5 per cent cap on oil sands production *after* GCOS won approval from the Conservation Board. This proved beyond any reasonable doubt that it was the government's intent to prove the commercial feasibility of the oil sands using the smallest possible plant. The government's wish to achieve development could only be realized by approving the present, amended GCOS proposal "which experience has shown to be the minimum possible size." [54] Evidently, the Board concurred with this logic and granted the request.

RESPONDING TO THE OIL SHALE THREAT

Thus, the province's basic oil sands strategy – which Manning expressed in 1962 and the Conservation Board interpreted in 1963 – guided development of the oil sands for the next five years. The Oil Sands Development Policy provided Great Canadian Oil Sands with the opportunity to continue with development free of direct competition. The conventional industry, so long opposed to the development of a synthetic industry, now seemed resigned to its new competitor. But in 1964, oil sands producers began to cast wary glances to developments in Colorado. American investors began seriously examining that state's oil shale deposits and a proposed shale plant expected to begin production around the same that the GCOS plant started up. [55]

Observers initially focused on North America's good fortune to have two unconventional deposits developed simultaneously. However, it did not take industry insiders long to regard shale oil as the single greatest threat to the future of the oil sands. Shale had the advantage of being in the United States and benefiting by its proximity to the major oil distribution hubs and markets. Moreover, oil companies looking to develop synthetic fuels in the United States would only have to contend with one tier of pro-rationing instead of the two confronted by producers in Canada. Developing Colorado shale no longer meant battling the Province of Alberta to increase the oil sands' production quota and then looking to gain a share of the restricted export market. Toronto Dominion Bank warned that once oil shale became competitive, it would be "a formidable competitor to any large-scale exploitation of the Canadian oil sands."[56]

The conflict between sand and shale simmered until early 1966 when Syncrude publicly urged the Manning government to allow other oil sands operations to go forward. In a speech before the Calgary Lions' Club, Frank Spragins, the president of Syncrude (the newly adopted name of the Cities Service, Imperial Oil, Royalite, and Atlantic-Richfield consortium), sounded an alarm. Syncrude's president warned the government that the industry was concerned with the activity, or lack thereof, in the oil sands. Unless the province went on with oil sands development, Canada's synthetic industry would lose out to the Colorado oil shale. Spragins acknowledged that additional oil sands development might harm the conventional industry, but any additional "unnecessary delay" might result in American shale oil surpassing the oil sands.[57]

A few days later, Opposition critic Mike Maccagno rose in the Alberta Legislature and stated that the production quota of 45,000 barrels per day set for GCOS might be too low. GCOS had revealed the previous October that capital costs for the construction of the plant were approaching $230 million, $40 million over original estimates.[58] Company representatives said they might need to increase their daily production total by an additional 20,000 barrels to 65,000 barrels per day to offset higher costs.[59]

Manning summoned industry representatives to a confidential meeting in Edmonton on May 11, 1966 to consider the revision of the province's oil sands development policy. Representatives from Syncrude, Imperial Oil, Atlantic-Richfield, Cities Service, Royalite, Shell Oil, Dome Petroleum, the Canadian Petroleum Association, the Independent Petroleum Association, and the Oil and Gas Conservation Board met with the premier behind closed doors. All participants remained tight-lipped about the results of the meeting to the press, but some details escaped. At the end of the year, *The Financial Post* reported the "strong possibility" that Manning was considering altering the province's oil sands development policy. Edmonton might increase the production allocated to the oil sands.[60]

The May meeting signalled a turning point for the province's Oil Sands Development Policy because so many of the oil industry's potential synthetic producers demanded change. Syncrude went into the meeting with guns blazing, and presented the government with a written submission that suggested ways of changing the development policy. "Within the past six months there have been many signs of weakness in U.S. domestic crude oil supplies," argued Syncrude. Syncrude contended that the American oil industry could not keep up with demand. All the states, except Louisiana and Texas, were already producing at or near capacity.[61] They calculated that supplies to fill future continental requirements were "in the hundreds of billions of barrels." Alberta oil sands and Colorado shales offered the only two major sources of supply meeting these requirements, claimed Syncrude. "The source of supply which can be developed most rapidly and economically will certainly enjoy the most favourable supply position." American companies, operating with government assistance, were driving ahead with development to shut the oil sands out of the American market altogether.[62] Whether the province acknowledged it or not, shale oil was now in a race with the oil sands. The confines of the provincial development policy might end up strangling the very industry it hoped to nurture.

If the Conservation Board and the Province hoped industry would provide solutions, they were disappointed. On this issue, the industry did not agree and no clear consensus emerged from the discussions. Shell argued that the present government policy did not present a "realistic opportunity" for oil sands development to take place. Commercial development of a successful oil sand operation depended on producing synthetic fuel in volume, something the present policy discouraged. IPAC and Dome Petroleum sharply disagreed with the idea of changing the development policy in any way, shape, or form. Changing the policy and increasing the production amount of synthetic fuel could "destroy the market incentive to search for and develop new conventional oil reserves in Alberta." The only way to distribute the risks equally would be to subject the synthetic industry to the same pro-rationing system encountered by the conventional industry.[63]

Meanwhile, the Canadian Petroleum Association refrained from offering any advice at all. Before the meeting, the CPA issued a survey to its members asking for their opinions about the changes, if any, that they wanted to the government policy. The CPA pledged that it would collect the results, assess them, and incorporate the findings into a formal submission to the Conservation Board. However, the results defied any easy generalization. While the majority clearly said that they preferred to keep the present policy intact, the overall results discouraged the Association from arguing one way or another.[64]

From the Conservation Board's point of view, the Oil Sands Development Policy clearly needed some changes. The 1962 decision stated that, as its major objective, the policy wanted to encourage the commercial development of the oil sands without unduly harming the conventional industry. The Conservation Board estimated that the current policy had accomplished the first objective but failed to satisfy the other. In fact, the current policy might delay the orderly development of the oil sands for the next twenty to thirty years.[65]

However, as the Conservation Board argued, justifying changes to the development policy would not be easy. All the available data pointed to the conclusion that increasing the capacity of the oil sands would doom the province's conventional industry. In 1964, the province reassessed Alberta's reserves and began a new pro-rationing plan for the conventional industry. The new plan included increased incentives to companies willing to begin enhanced recovery operations in producing fields. Many took advantage of the new provisions, and the use of enhanced recovery techniques led to higher crude oil recoveries in older fields. Meanwhile, the province's reserves received an unexpected shot in the arm when new exploration techniques also added substantially to provincial reserves. In 1964 and 1965, two large discoveries in the province increased the life index of Alberta's crude reserves from twenty-two to thirty-one years. Concurrently, the shut-in capacity of the conventional industry continued to grow to 60 per cent by 1965.[66] The potential cost of delaying oil sands development any longer, however, might destroy the industry. George Govier, chair of the Conservation Board, suggested that tinkering with certain elements of the 1962 policy might enable the other projects to continue.[67]

Both Manning and the Conservation Board understood that without adding substantially to Alberta's crude markets, justifying any increase in oil sands production would be difficult. While the Japanese Government initiated exploratory discussions with the province about purchasing additional production from the oil sands, negotiations dragged on for much of the mid- to late-1960s without agreement.[68] Unless the province opened the throttle on development, the oil sands might continue to lose ground to shale oil.

Manning instructed the Conservation Board to continue its consultations with industry in the hope of finding some common ground. However, as discussions between industry and government continued through 1967, no clear consensus emerged. The Canadian Petroleum Association encouraged the province to expand oil sands production while the Independent Petroleum Association bitterly opposed any alteration to the current development plan.[69] IPAC maintained that a barrel of synthetic crude would displace a barrel of conventional crude. Given the current state of Alberta's shut-in capacity, it made little sense to increase production unless an alternative market, like Japan, was willing to take oil from the oil sands.

In any case, IPAC also reminded the Province that the economics of the oil patch had not changed materially since 1962 when it intervened in the GCOS hearing. For every barrel of conventional oil produced in the province, the government received $0.95 in lease bonus payments, fees, rentals and royalties – equal to forty cents out of every dollar of oil sales. Synthetic crude, on the other hand, only provided $0.30 in lease bonus payments, fees, rentals and royalties – or some twelve cents on every dollar of synthetic sales. For each barrel of oil the province added to the oil sands quota, the province would lose $99.68 a year.[70]

While IPAC's argument made perfect sense in the short term, Manning and his Cabinet were more concerned with the long-term implications of the province's development policy. The presence of the world's single largest deposit of oil sand meant nothing if the industry ignored the sands and pursued oil shale development. After a year and a half of stalled negotiations, Manning sought to ease restrictions on oil sands development. He encouraged the Syncrude group to resubmit its proposal, regardless of the consequences for the conventional industry.

So, in February 1968, Manning presented an amended oil sands development policy in the Alberta Legislature. The premier acknowledged the conventional industry's difficulties but countered that the oil sand development policy needed revision to "encourage greater market growth than would otherwise occur." Crucial to the government's argument was its contention that growing supply deficiencies in the United States would require increased imports of Canadian oil. Therefore, increased development of the oil sands would establish Alberta's reputation as a plentiful source of supply. Furthermore, development would enable the oil sands to maintain its technological advantage over the American oil shale.[71]

As much as possible, the new policy attempted to unshackle progress on the oil sands. It implicitly acknowledged that government policy – not industry practices – had limited the development of Canada's synthetic fuels industry. The first part of the policy focused on increasing the share of the market available to synthetic fuels by clarifying the distinction between "within reach" and "beyond reach" markets (Table 5.1). "Beyond reach" markets included any market that Alberta's conventional industry was not currently serving or could expect to serve in the future because of price, quality specifications, or other reasons. If a producer could show that they would be serving a "beyond reach" market, the province would place no restriction on production. Meanwhile, where the applicant's proposal would provide growth by the development of a new "within reach" market, the province would permit production from the oil sands in volumes equal to 50 per cent of that market. However, the total volume of commercial oil sands production, including that contained in projects already authorized, allowed to enter new "within reach" markets would be limited to a *total* of 150,000 barrels per day for five years.[72] A project proposing the marketing of oil sands production in a "within reach," but not new, market, would gain approval when the life-index of the conventional industry made development possible. To encourage further research and development, the province declared that experimental operations "may involve temporary production and marketing at sub-commercial levels." Finally, there would be no change to the policy regarding "beyond reach" markets. Any project seeking to serve these areas would not be subject to pro-rationing.[73]

The new policy altered the Conservation Board's approach to calculating the supply and demand for Alberta crude oil, freeing room in the province's pro-rationing scheme to increase oil sands production. In the first part of the decade, the Board devoted considerable attention to calculating the total level of *Canadian* crude production. It then determined the proportion of the market served by Alberta's producers. In 1963, this allowed the OGCB

Table 5.1   Alberta Conservation Board's Market Definitions and
            Allowable Production Levels, 1968

| Beyond Reach Market | | Within Reach Market | |
Geographically Removed	Specialty Markets	"New"	Other than "New"
No Restriction	No Restriction	Total "New" market to double the synthetic crude production must be provided. Maximum approvals under this criterion limited to 105,000 barrels/day over 5 years	Volumes as desirable to ensure the life index for conventional oil does not fall below 12–13 years

Source: Oil and Gas Conservation Board Report 68-C, pp. A4-A15.

to justify turning down the Cities Service and Shell applications because of anticipated production from new fields in the Northwest Territories. The new policy, however, instructed the Conservation Board to only consider *Alberta's* projected supply of crude oil between 1970 and 1980. Since the Conservation Board believed Alberta's conventional industry would reach its capacity sometime in the following decade, the need for additional production from the oil sands seemed justified.[74]

From the province's point of view, the new policy represented the best of all possible worlds. Commercial development of the oil sands would continue and the government could control growth. Rather than releasing a blanket policy, the development policy enabled the provincial government to choose which projects it would approve. The province, through the Conservation Board, would evaluate each proposal on a case-by-case basis and act on those projects it believed had the best chance of succeeding.

Thus, while the province increased the net amount of oil sands production, it only intended to allow a fraction of that portion to reach the market in the near future.

The subtleties of this arrangement were not immediately apparent to all segments of Alberta's petroleum industry. Several conventional producers worried about the new emphasis on synthetic fuels. Because the Canadian Petroleum Association chose to remain silent on the issue, an informal coalition emerged. Many large firms and the Independent Petroleum Association of Canada came together and voiced their opposition to the provincial plan.[75] IPAC objected vehemently to the new policy and issued a strongly worded statement in early March. It argued that the new policy was not in the province's interest. Assuming that the U.S. would increase imports of Canadian oil was foolhardy. No one could accurately predict what direction the quota would move. Meanwhile, government claims that the new policy would maintain the oil sands competitiveness against the oil shale seemed questionable. "The Association is convinced that development of techniques in the U.S. will proceed independently of and unaffected by Canadian developments." If anything, IPAC argued that the new policy "might well induce accelerated efforts on the part of the U.S. in this regard."[76]

THE SYNCRUDE APPLICATION

IPAC's concerns had little impact on the provincial government and, with the new policy in place, the province removed the last remaining political obstacle to the Syncrude group. On May 3, 1968, the group submitted its revised application to the Conservation Board. Instead of seeking the 100,000-barrel per day facility outlined in the 1963 application, Syncrude proposed the construction of a smaller facility – 80,000 barrels per day.[77] Catering to the government's new definitions of "new within reach" markets, the consortium informed the Conservation Board that the United States would absorb all of the facilities' production. The claim of an ever-expanding U.S. export market generated a good deal of speculation and comment by the Conservation Board and IPAC. The Conservation Board believed that the life index of the U.S. market would continue to decline as it had every year since 1958. Meanwhile, it also projected that decreases in the Canadian supply would permit an increase in oil sands production.

On July 5, IPAC Vice-president Charles S. Lee wrote to members of the Association's National Oil Committee and informed them of a significant change in the IPAC's strategy. Until now, many of IPAC's statements on the Oil Sands Development Policy and the Syncrude application opposed development for "public interest" reasons – the impact of synthetic crude development on government revenues and the conventional industry. Now, Lee argued that the only way the Association could win converts to its cause would be by "establishing the uncertainties of interpretation of present government policy."[78]

In many ways, the change in strategy signalled the recognition on IPAC's part that the Syncrude application would probably succeed in spite of their best efforts. Dark stories circulated throughout IPAC that after the crucial May 1966 meeting between Manning and representatives of the industry, the latter threatened to pull themselves – and their considerable assets – out of the province unless the government modified its position on the oil sands. "The large sums of prime risk capital supplied to Alberta by the major oil firms over the last 20 years – their tremendous control of refineries and markets throughout North America was undoubtedly a decisive factor in formulating the new policy."[79] Ultimately, IPAC concluded that the province had already determined that it would increase oil sands production. "There is simply too much at stake for the Alberta government to dispose of [the application] in any other way," concluded one IPAC member. "Mr. Manning, in the final analysis, will assume that provisions included in his oil sands policy will prevent any serious dislocation on the part of the conventional industry." IPAC believed that "after a rather stormy hearing," Syncrude's application would receive approval.[80]

IPAC hoped that by pointing out enough contradictions between the Syncrude application and official government policy, it might win some concessions. Perhaps these would mitigate the impact of the oil sands on the conventional industry. Thus, IPAC worked diligently to bring these contradictions to light. In a press statement, IPAC launched an all-out assault on the Oil Sands Development Policy and the Syncrude application.

The Department of the Interior had just released a ruling indicating that they would treat synthetic crude no differently than Canada's conventional crude. The announcement buttressed IPAC's claim that additional synthetic production would only displace conventional crude.[81]

Furthermore, IPAC announced that it did not agree with the province's definitions of "within reach" and "beyond reach" markets. The new policy allowed the applicant to determine what constituted such markets. This meant that Cities Service's pledge to take its portion of production to its American refineries enabled Syncrude to claim it would service a "new within reach" market. In any case, IPAC bitterly retorted that few major markets existed on the North American Continent that were not physically or practically connected "by pipeline or seaway to Canadian crude." The press statement concluded with the suggestion that, at the very least, the conventional industry should expect some marketing concessions from the government. Rather than granting synthetic crude equal access with conventional crude to "new within reach" markets, they should allow conventional producers to provide two barrels for every barrel of synthetic crude.[82]

Despite IPAC's considerable efforts, the Association feared that it would wind up with very few concessions from the Conservation Board. Not even a negative report, which IPAC commissioned from internationally renowned petroleum consultant Walter J. Levy, appeared to dissuade the Conservation Board.[83] The Board's final report on the application supported the applicant in every respect. The Board rejected IPAC's assertion that the United States would not alter its import quotas in Canada's favour. It perceived that North American security interests would prevail.[84] Syncrude's application, declared the Board, was "satisfactory in all respects except for the marketing plan." The Board thought Syncrude had not accounted for the "probable magnitude and rate of development of the recent Alaskan discoveries." So the Board delayed approval of the application until it convened a public hearing to assess the impact of Prudhoe Bay on the need for Alberta oil sands.[85]

Infuriated, Syncrude appealed directly to the Provincial Cabinet in February 1969 to overturn the Conservation Board's decision. Although Syncrude understood the Board's concerns, they argued that any additional hearing would "not be of decisive assistance to the Board." Syncrude did not want to make a forecast of Alaskan reserves until it had more solid information. "These evaluations will only evolve over the next few years, and in order to be meaningful such evaluations would have to be based upon events which will occur step by step during this period."[86]

Unstated, but present in Syncrude's objection, was the fact that the Board's decision placed Syncrude in a difficult situation. If Syncrude underestimated the size of the Alaskan reserves, the actual figures could force the Conservation Board to delay Syncrude's approval in the future. If, on the other hand, it overestimated the size of the reserves, it might enable the Conservation Board to delay the application even longer. Syncrude concluded its submission by pointing out that it had been waiting for a final decision since 1962. Further delays made it exceedingly difficult for the four partners to justify their continued support for the oil sands. The company believed it was entitled to expect "a prompt decision" from the government.[87]

To its credit, the province did not overturn the Board's decision, and informed Syncrude that it would have to go before the Conservation Board again. On May 26 and 27, 1969, the Conservation Board opened hearings on the Syncrude application for the third time. The Board only conducted hearings on Syncrude's proposed marketing plan since their December 1968 report had deemed the rest of the proposal satisfactory. After the drama of the preceding years and months, the hearings were anticlimactic. The Conservation Board realized that if they did not approve the Syncrude application, "it may be several years before another application is made and further commercial development could be delayed well beyond 1976–77."[88]

The Board issued a split decision in September 1969, two to one, in favour of allowing the application to proceed. In the majority opinion, written by George Govier, the Board declared itself satisfied with Syncrude's marketing plan. Anticipated deficiencies in the United States supply by 1976–77 suggested the elimination of restrictions on the import of Canadian oil by 1977 at the latest. Even if the Board's estimates were wrong, Govier estimated that the increased oil sands production would only defer 2 or 3 per cent of the market for Canadian oil.[89]

IPAC had anticipated this outcome a year earlier. With no additional markets for Canadian crude available, some conventional producers – particularly the independents – resigned themselves to the belief that the province would never allow them to be competitive again. Although it had taken nearly twenty years to complete, the Province of Alberta finally devised and implemented an oil sands development strategy. The energy crises of the 1970s would test the wisdom and effectiveness of this effort.

CHAPTER 6

Flexibility and Paralysis: The Oil Shocks, Government Policy and Intervention, 1970–77

I'm the Sheik of Cal-gary

These sands belong to me

Trudeau says they're for all

Into my tent he'll crawl

Like Algeria did it to DeGaulle

The gas we've got today

We just don't fart away

Gas pains don't worry me

Cuz I'm the Sheik of Cal-gary

    – Satiric Song, 1974[1]

The current lack of major oil discoveries in Canadian frontier areas precludes the possibility of significant production from those areas before the mid-1980s. Moreover, if exploration in frontier areas fails to achieve the desired results during the next several years … it is apparent that self-sufficiency in the 1980s is an unattainable goal without considerable oil sands production. – *Canadian Petroleum Association, June 5, 1975*[2]

The oil shocks of the 1970s radically altered political attitudes on both sides of the Canadian-American border. As governments groped for solutions to the energy crises of the 1970s, some advocated "quick fixes" to long-term problems. Schemes that emphasized nuclear reactors, solar power, biomass, grain alcohol, synthetic fuel, and countless other programs, came to the attention of scholars, business leaders, and governments. But in real world circumstances, most of the quick and simple solutions proved impractical or were impossible to implement. In the short run, they could do little to stabilize oil prices, or to break the OPEC cartel's stranglehold. As Saudi Arabia supplanted the United States as the world's "swing producer," the ability of the United States to influence world oil prices waned appreciably. A series of administrations searched for new solutions. Richard Vietor, a Harvard economist, concluded that, after November 1973, the Nixon administration had already decided to emphasize synthetic fuels development. "The policy at issue was not whether government should subsidize development of synfuels, but rather, how fast and in what manner," wrote Vietor.[3] As Aramco president Frank Jungers bluntly told the Saudi Government, OPEC's actions kick-started the alternative energy industry and the energy conservation movement. "[OPEC] had simply made oil too costly, and people naturally sought other alternatives."[4]

After the 1973 oil embargo, significant changes took place in political and commercial circles. The psychological effect of the embargo – the threat of rapidly diminishing supplies, gas lines, and chronic shortages – succeeded in heightening market concerns about supplies. Prices on the spot market went higher as panicked buyers snapped up whatever oil they could find.[5] The crisis mentality of the market, hoarding, and skyrocketing crude oil prices, simply encouraged producers to squeeze the market even more. Keeping oil in the ground became synonymous with cautious investment for the future as analysts produced doomsday scenarios with crude oil prices reaching $100 a barrel. This outlook extended beyond OPEC to other oil producing countries, including Canada, and brought greater government intervention in the oil industry. Higher energy prices prompted Canadian Prime Minister Pierre Trudeau to try to change the continental orientation of the oil industry to a national direction. Ottawa proclaimed that the "national interest" required Canada to be energy self-sufficient by the 1980s and Trudeau looked at the oil sands with renewed interest.

However, the energy crisis also brought new problems for policymakers and businessmen alike concerned with the growth and development of Canada's synthetic fuels. On the one hand, OPEC's actions helped make the oil sands a viable alternative source. Higher world crude prices meant that synthetic fuel projects were profitable in more North American markets. On the other hand, the more profitable synthetic fuel projects became, the more likely they would become targets for government regulation. The federal government began to embrace a much more vigorous nationalism as its touchstone. Suddenly, the oil sands became a federal priority, much to the consternation of Alberta Premier Peter Lougheed who regarded energy issues as the exclusive domain of the province.

For Canada's oil sands producers, the beginning of the 1970s offered some encouragement that synthetic development would soon be an important part of the Canadian oil industry. Increased international demand for crude oil and shrinking conventional reserves seemed to suggest the possibility of unfettered access to world markets in the future. However, the sobering reality was that additional commercial facilities lacked guaranteed sources of revenue. The experience of Great Canadian Oil Sands during its first decade of operation served as a particularly vivid cautionary tale. From the very beginning, GCOS encountered the usual start-up problems and a few new ones as well. Besides building the first major commercial oil sands production plant, the company had to build a twenty-mile road to the site, construct a $3.3 million bridge spanning the Athabasca River, and lay a 266-mile pipeline to Edmonton to bring the bitumen to the refinery. Despite the fanfare that greeted the initial public offering of GCOS shares in 1964, a 1967 audit of the project estimated that Sun Oil furnished 99 per cent of the $240 million to bring GCOS into operation.[6]

Given the high construction costs associated with the plant's start-up, Sun Oil and GCOS officials had both looked forward to the beginning of operations in 1968.[7] Although the plant design could produce 45,000 barrels per day, technical difficulties and delays resulted in daily production of only about 24,000 barrels per day in 1968. The company had anticipated

Table 6.1   GCOS Operations, 1968–78

Year	Revenues	Profits (Losses)	Allowable Daily Production (barrels/day)	Avg. Actual Daily Production (barrels/day)
1968	6,370,269	(8,667,641)	45,000	15,000
1969	28,471,961	(25,546,111)	45,000	27,300
1970	34,738,961	(16,028,033)	45,000	32,700
1971	50,724,000	(8,246,000)	45,000	42,200
1972	62,284,000	(680,000)	45,000	50,900
1973	73,000,000	(2,300,000)	45,000	41,587
1974	108,000,000	4,000,000	65,000	45,700
1975	124,434,000	(991,000)	65,000	42,555
1976	159,827,000	11,962,000	65,000	47,750
1977	180,558,000	12,917,000	65,000	44,914
1978	213,186,000	19,012,000	65,000	44,750
Total	1,041,594,191	(23,405,683)	55,000	43,536

Source: Great Canadian Oil Sands *Annual Report* for given year.

these initial problems to a certain degree. It had estimated that it would take Great Canadian approximately one year before its facilities could produce approximately 37,000 barrels per day. But, during the second year of operations, the plant fell far short of company expectations. Rather than producing the desired 37,000 barrels per day, actual production barely exceeded 27,000 barrels per day (Table 6.1).[8]

Tar sands are very thick, and the oil flows very slowly – if at all – under normal operating conditions. GCOS soon realized it would be involved in an incredibly cumbersome, and painfully slow, process when temperatures plunged to between –20°C and –40°C in winter. The machinery used to move the tar sands to the refining facility often broke down. The huge bucket wheels used to move the sands to the refinery aged visibly from the sheer effort and exposure to harsh conditions. Steam traps crystallized the atmosphere around the plant to such a degree that workers could see only two to three feet.[9] One major challenge was keeping an adequate work force on hand. Despite free room and board and double pay for overtime, GCOS experienced an entire turnover of personnel every 2.7 months. "Recruits were brought from Britain, Cuba and even as far away as Korea," noted one early article on GCOS. "The chief pinch occurred in such skilled trades as pipe fitters, electricians and welders."[10]

In spite of Sun's continued investment, modifications, and attempts to improve the facility, GCOS amounted to little more than a vast money pit. Between 1963 and 1970, Sun Oil invested more than $250 million in the oil sands and the company was still losing money. In late February 1970 H. Robert Sharbaugh, the president of Sun Oil, and George Dunlop, Sun's Chairman of the Board, met to discuss the company's plans with the oil sands. "Correctively measures to enhance the financial viability of this project must be immediately undertaken," Sharbaugh insisted. It was with more than a measure of frustration that Sun implored the provincial government to re-examine its royalty regime. "It is mandatory that our cash losses be stopped," Dunlop advised the new Alberta Premier Harry Strom.[11]

In a series of face-to-face meetings between GCOS officials and the Alberta government, the company suggested that instituting a flat 8 per cent royalty would greatly improve the plant's situation. The province concurred with GCOS's evaluation, and on May 26, 1970 reduced GCOS's crown royalty 50 per cent (from 16 per cent to 8 per cent) for the next three years. The altered royalty schedule trimmed $9.5 million from GCOS's losses and signalled to the industry at large that the province remained committed to the development of the oil sands. Significantly, however, the province did not formally alter the royalty provisions of the Mines and Minerals Act, nor did it change the Oil Sands Development Policy. Privately, the provincial government suggested that it was at least considering altering the development policy, but put nothing on the table for the industry to consider. Vague promises that "something" would change circulated throughout the industry. For other companies interested in pursuing commercial development of the sands, the moves – or lack thereof – were cause for concern. How could synthetic fuel projects get off the ground if the province insisted on enforcing an unrealistic royalty schedule?

Syncrude's troubled beginning foreshadowed problems with similar projects. Although the OGCB approved the Syncrude project in September 1969, rising labour and construction costs chipped away at potential earnings. To offset rising costs and to help make the project more attractive to outside investors, Syncrude applied to the Conservation Board in 1971 to raise production from 80,000 barrels per day to 125,000 barrels per day.[12] To support its application, the company noted that consumer demand finally caught up with supplies, ending the twenty-two years of surplus in North American production. Combined with the growth in Canadian exports to the United States – they would total one million barrels per day in 1972 – Syncrude's increased capacity was needed to offset other production losses.[13]

Developments south of the border supported Syncrude's conclusions. Three years earlier, in 1968, the U.S. State Department notified foreign governments that production from the American oil industry would soon reach the limits of its capacity. In 1970, America's domestic crude production peaked and began to decline. Gradually, imported crude from the vast oil fields of Saudi Arabia replaced those of Texas as the crucial "swing producer" whose production determined the world price of crude oil. Combined with opportunism of the OPEC cartel and the beginnings of the nationalization movement in the Middle East, the shift marked an important turning point in world crude markets that few appreciated.[14]

With demand for Canadian crude still growing, North America would clearly become even more dependent on imports to meet the growing discrepancy between supply and demand. Ottawa continued to press Washington about expanding Canada's share of the U.S. market and belatedly suggested formalizing the relationship with a limited continental energy agreement. Ironically, in the pre-embargo days, Canadians were more interested in such a deal than their American counterparts. Discussions about a continental energy policy began in 1968 shortly after the discovery of natural gas deposits in Alaska. Since the Canadian North contained the same geological formations, Ottawa argued that both countries should pursue the development of a northern pipeline to ship natural gas to the continental United States. A U.S. cabinet task force, headed by Secretary of Labour George Shultz, investigated the possibility and concluded that a formal continental policy was desirable but faced an uphill battle. To be effective, Shultz wrote that the United States should try to wean Eastern Canada from its dependence on foreign imports. Any common energy policy would have to divert surplus Alberta oil to eastern Canada to seal the market from foreign imports. The United States would trade access to "the high price[s] of the U.S. market" in exchange for "definite assurance in times of emergency that the flow of Canadian oil continues."[15]

Pragmatic politics, however, prevented Ottawa from reaching an accord. Quebec, the most politically volatile of the provinces, had become dependent on cheap imported oil and Ottawa was reluctant to upset this relationship. The federal government feared that any move to deny French-speaking Canada inexpensive energy might encourage the growing separatist movement. Washington replied that the Canadian government ignored "adequate national security precautions" as it adhered to a national oil policy that treated eastern Canada differently from the West.[16] Since Ottawa refused to budge on integrating the entire Canadian market, there was little point in discussing the integration of the North American market.

Negotiations on a common oil policy continued to lurch along mostly because Ottawa desperately wanted to expand the Canadian industry's access to the American market without compromising the domestic provisions of Canada's oil policy. To sweeten the deal, and draw attention away from the two-price system, Ottawa held out the riches of the oil sands to the Nixon administration. If the United States accepted the Canadian proposal, the proposed Alaskan pipeline could hold production from the oil sands. Such a pipeline would "go a long way toward making production of oil from the Athabasca Tar Sands economical" and promise the United States "an almost limitless supply of oil."[17] In the final analysis, although the idea of a common energy policy intrigued elected officials on both sides of the border, neither side could give enough in negotiations to reach a final agreement.

Meanwhile, Canadian producers began wondering where the oil to meet growing domestic demand would come from and inadvertently touched off a political maelstrom in the process. In 1970, the CPA prepared an estimate of Canadian reserves using dubious methodology – the CPA estimated the country's total cubic mileage of sedimentary rock and then estimated the oil and gas within each basin by comparing them to similar basins in the United States. The result was wildly inflated figures of Canadian reserves that were then used to justify continued oil exports to the United States. "At 1970 rates of production," noted federal Energy Minister Joe Greene in June 1971, "[Canada's] reserves represent 923 years of supply for oil and 392 years for gas."[18]

A month earlier, however, the Canadian Petroleum Association began to review the 1968 Oil Sands Development Policy at the request of the provincial government and established an Oil Sands and Heavy Oil Committee to direct this study. Understandably, the CPA wanted to examine the oil sands development policy within the broader context of the oil industry at large. The Economics Study Committee prepared an estimate of the demand for and supply of Canadian crude oil until 1980. The results of the Economics Study Committee sent shock waves throughout the Association and the country. Instead of using the same method the CPA used a year before, the organization altered its criteria. The result was a pessimistic forecast showing a rapid decline in the life index from 24.2 years in 1970 to 11.2 years in 1975.[19] With the long construction times and huge capital investment required to bring a commercial oil sands plant on stream, the CPA now claimed the province faced a dire situation. "It is not now possible to develop supplemental production from the oil sands to prevent the decline in Alberta's life index below the 12 to 13 years of supply," the industry concluded.[20]

From the CPA's perspective, Canada clearly faced supply disruptions unless Alberta quickly adopted a policy supportive of oil sands development.

The CPA's Oil Sands Committee identified crown royalties as the most critical area for the province to address in any re-evaluation of the policy. When the provincial government first outlined its royalty structure in 1962, it pegged the royalty to the cost of a finished barrel of oil. As mentioned above, from GCOS's point of view, this effectively served as a disincentive to production and stood in marked contrast to the province's other taxation schemes. The government did not apply taxes to the value of the raw material, the bitumen. Instead, GCOS and other producers argued that the government was unfairly taxing the finished product, the synthetic oil, to which significant value had been added through refining and upgrading. Beyond this consideration, GCOS's experience demonstrated that the economic viability of any future commercial oil sands projects depended on the provincial royalty regime.

A NEW PROVINCIAL GOVERNMENT

Provincial politics intervened before the policy review was completed. The Social Credit government called an election for August 31, 1971. During the campaign, critics characterized Premier Harry Strom's government as "burnt-out" and on election day voters emphatically rejected the party. Voters chose Peter Lougheed's Progressive Conservative Party to form the next government. The change in ruling parties was the first for Alberta since 1935. Although some pundits believed the difference between the two parties was more style than substance, it soon became apparent that not all changes were cosmetic.

In part, Lougheed's changes to provincial oil policies mimic the conditions outlined by Harvard economist Raymond Vernon's theory of the "obsolescing bargain." Vernon's theory posits that a host government possessing valuable natural resources it cannot exploit on its own because of a lack of capital, technology, or marketing power is in a poor bargaining position *vis-à-vis* a multinational corporation (MNC). Since the MNC possesses the assets, skills, or technology required, a host government must offer generous concessions (i.e. lower taxes, deferred royalty payments, etc.) to ensure that the MNC will undertake the risk of developing the resource.

However, Vernon argues that the bargain struck between a government and a company gradually becomes obsolete; the demand placed on the social services (the creation and maintenance of infrastructure, education, health care, etc.) of the host government grows exponentially while its net revenues increase marginally. Particularly after the venture is operating successfully, the perceived level of risk undertaken by the MNC diminishes and the host government seeks to revise the terms of bargain.[21]

Although the policies of Lougheed's government still satisfied the province's business interests, he nevertheless regarded the oil industry with suspicion.[22] "Lougheed never seemed to trust the big oil companies," wrote journalist Peter Foster. "His objection was the politically obvious one that their interests were not the interests of Alberta. They could pull up and move out any time prospects looked better elsewhere."[23] Moreover, the newcomers believed that the Social Credit government had coddled and protected the oil industry during its infancy and, as a result, the industry grew accustomed to its preferential treatment. Lougheed warned that the industry should no longer expect the same level of deference from the provincial government and promptly announced an increase in royalty payments on conventional oil. To the Syncrude group, Lougheed insisted that the company allow Albertans to buy shares in the company, that they appoint a Canadian director to oversee operations, and that "wherever possible" they use Alberta technology to build the plant.[24]

Peter Lougheed and the Conservatives also affected the commercial development of the oil sands. The personal relationship between Ernest Manning and various oil company executives influenced the development of the oil sands during the Social Credit years.[25] This does not imply that the interpersonal elements were not important during the Lougheed era – they were then and remain so today. But while formalizing government practices, the Conservatives imposed a bureaucratic process on the oil industry to a degree not previously experienced.

Previously, the province and oil sands developers settled contentious issues informally. After Lougheed took power, industry insiders found themselves dealing with committees of civil servants, many of whom were new to their jobs. During his first year in office, the new premier replaced more than 70 per cent of Alberta's senior civil servants.[26] As a result, the oil industry could not take for granted that it was speaking to a government that knew the details of commercial oil sands development in the province. In an instant, the Lougheed government swept away the institutional knowledge of the industry built during the Social Credit years. The CPA now found itself explaining basic operating principles to inexperienced leaders.

Gone too was the sense of humility evident during Manning's tenure in office. Nearly fifty years had passed since Alberta began development of the oil sands. The Social Credit government held office for thirty-six of those years. Repeatedly, experience and bitter lessons taught the Social Credit government not to expect instant solutions from the oil sands. In the twenty years since the Blair Report, the province constructed only one synthetic fuel plant. The Conservatives had no such modesty. Despite GCOS's continued difficulties, the Lougheed government believed it possible to construct between ten and twenty commercial oil sands facilities within five years.[27] Lougheed's Conservatives possessed an unswerving faith in the ability of science and industry to triumph over nature. Edmonton's message was clear: if the pace of oil sands development over the past twenty years was slow, it was because the companies owning oil sands leases did nothing with them. Lougheed's government would not tolerate such dawdling any longer.

Discussions between the province and industry in May 1974 illustrated the new dynamic in provincial policymaking. To nudge producers along, the province established the Oil Sands Study Group to consult with industry and guide provincial policymaking. In relatively short order, the Provincial Oil Sands Study Group proposed many changes to the lease provisions of the Oil Sands Development Policy. The time seemed opportune as the original twenty-one-year leases signed by the Social Credit government would expire within the next three years. The new government wanted to make a policy that would speed development. The study group believed industry recalcitrance was the main reason oil sand development languished. It suggested development would increase if the province shortened lease terms and reduced the acreage offered to developers. Thus, the proposal suggested a fifteen-year lease and shrinking the current lease size of 50,000 acres to 25,000 acres.[28] The shorter term would encourage developers to begin work immediately while the new lease sizes would, presumably, grant other interested parties access to the sands. Doubling the number of available lease sections would also increase provincial revenues.

The CPA and IPAC patiently explained that it was the province, not the oil companies, that had dictated the pace and scale of oil sands development since the 1950s. In any case, the fifteen-year lease term did not provide enough time for companies to make the necessary preparations required. Commercial ventures needed time to "delineate reserves, develop a process suitable to his particular deposit, obtain the multitude of permits and design and construct a plant."[29] As for the smaller lease size, the submission simply pointed out that the quality of the bitumen deposit varied greatly throughout the section. Given the state of current technology, limiting lease sizes might result in less production from the sands rather than more. Smaller lease sizes would hamper development since the oil sands were dependent on scale for profits. Nevertheless, if the province insisted on pushing for the Oil Sands Study Group changes, that was its prerogative. But both the CPA and IPAC warned that industry would regard these items as detriments to additional production.[30] The province soon dropped the two controversial proposals, but the negotiations illustrate the industry's difficulties with the new provincial government.

While layers of government bureaucracy may have tried the patience of the oil industry, it learned to negotiate with the private sector. Less predictable, however, was Lougheed's relationship with the federal government under Prime Minister Pierre Trudeau. Both men were skilled politicians and gifted negotiators with strongly-held beliefs. The industry's great misfortune during the 1970s was that these two powerful figures frequently disagreed on a number of energy-related issues. Indeed, with the possible exception of Trudeau's struggle to patriate Canada's constitution, no issue received more attention, or was waged as ruthlessly, as the energy battles between Edmonton and Ottawa.

Part of Lougheed's toughness stemmed from his training as a lawyer. But Lougheed also believed that previous provincial governments were too meek when it came to negotiating with Ottawa. While a member of the Opposition before 1971, Lougheed and the Conservative energy critic, Don Getty, repeatedly criticized what they saw as the Social Credit government's passivity. Alberta, they pointed out, owned the resource but did not participate in the decision-making process.[31] After his party's victory at the polls in 1971, Lougheed vigorously pursued an expanded role for the province in national decision-making, particularly as it related to energy issues. At the National Press Club in Ottawa in November 1971, Lougheed referred to his government's new, more militant, policy. "If Alberta poker chips are involved at the poker table," declared Lougheed, "we will be at that table."[32]

Predictably, the Trudeau government dismissed Lougheed's statement out of hand. The province only had control over resources until they passed through a provincial border when they came under federal jurisdiction. For civil servants in Ottawa, it seemed Lougheed was engaging in political posturing. But when Lougheed unilaterally increased the price of Alberta oil and gas a few months later, Ottawa took notice. It was only a matter of time before they would join the battle.[33]

In early 1972, the CPA began lobbying the province to remove all restrictions on oil sands development. The association sought to address the anticipated shortfall between production and consumption before it became a serious issue. According to the CPA, a solution could be provided by streamlining the application procedure and encouraging oil sand development. The CPA believed these simple moves would attract new investors and stimulate production. Since obtaining permission to drill for conventional oil and gas was not necessary for firms, the CPA argued that oil sands producers should enjoy the same freedom.[34] "During the past 11 years the 1962 Oil Sands Development Policy and its subsequent amendments have either prohibited or severely restricted the holders of Oil Sands or Bituminous Sands Leases from obtaining building permits to build plants and to develop commercial production from these leases." The Canadian Petroleum Association now wanted these restrictions to end immediately.[35]

However, removing restrictions on oil sands development did not necessarily mean that commercial development would go forward. There were many places for investors to place their money in Alberta during the 1970s. A poorly-defined growth philosophy obliged the provincial government and businesses to choose between competing projects. For oil sands developers, this also meant that factors besides access to capital affected the pace of development. Critical shortages of available skilled labour, equipment, and materials drove operating costs higher as companies and industries openly competed with one another for resources. In the scramble for scarce capital, equipment, and personnel, oil sands producers laboured at a decided disadvantage. A critical lack of housing and infrastructure around Fort McMurray made it difficult to attract new employees, or to deliver material and equipment to the region. Long lead-times, combined with the uncertainty of wildly fluctuating production levels, worried investors. Indeed, after Syncrude's amended proposal received approval in 1971, the company began a concerted effort to complete conceptual engineering plans and to pin down estimated construction costs for the project. The studies took over a year to complete, and through their investigations, it became apparent that construction of the plant would create a shortage of skilled labour in the province. The problems identified by Syncrude's report were serious enough to merit special attention by the CPA and both levels of government.[36]

But even before discussions between industry and government began, the industry knew it faced an uphill battle. Removed from daily considerations since Abasand's failure, both Ottawa and the National Energy Board struggled to get up to speed on even the most basic issues. The CPA told the federal government that private firms faced several daunting obstacles in developing the oil sands. First, as late as 1972, only surface-mining techniques had proved viable despite extensive research on *in situ* methods. Second, the capital investment required for simply *processing* each barrel of oil produced from the sands averaged between $4,000 and $5,000.[37] A plant capable of producing between 100,000 and 125,000 barrels per day would cost $500 million to build. Third, inflation made it difficult to estimate what the final construction, supply, and labour costs would be. Finally, synthetic producers warned the federal government about the long lead-time needed to address any shortfall in Canadian supplies. Even the industry's best-case scenarios suggested that it would take at least five years to construct a new oil sand plant and begin production.[38]

In addition, the changing domestic political scene began to exert more influence over provincial and federal politicians, but in very subtle ways. The Conservative government in Alberta understood the importance the industry attached to a new oil sands development policy, but delayed a policy announcement nonetheless. From Edmonton's point of view, the rapidly changing world situation counselled patience; Lougheed's government feared that if it committed itself to a long-term policy, the volatile nature of the international market would leave the province with inadequate revenues. Although the provincial government continued to consult with industry about a new strategy, it steadfastly refused to commit to a long-term development policy. Instead, it preferred to remain flexible, particularly since negotiations with Syncrude continued to drag on.

The stance frustrated the industry, but satisfied the provincial government's purposes. For the Conservative government, granting approval on a case-by-case basis enabled them to exert some discretion over what was essentially Social Credit policy. The province could also ill-afford another plant beset with problems like GCOS. "As far as the international financial community is concerned," Lougheed notified the Legislature, "the [GCOS] project, to put it mildly, was a financial setback. The second plant [Syncrude] must succeed."[39]

THE OIL SHOCKS

The oil shocks represented the culmination of two separate events in the world oil market – the "leapfrogging" oil prices begun by Libya in 1970 and OPEC's use of the oil "weapon" to link trade with developments in Middle Eastern foreign policy. The combination of forces drove international crude oil prices up from under $3.00 in 1972 to $10.50 in early 1974. In September 1973, Energy Minister Donald Macdonald froze the price Canadians paid for domestic petroleum products as part of Canada's response to rising international prices. Ottawa would subsidize the difference between the domestic price and international rates (Table 6.2).[40]

Increasing federal taxes on oil companies would fund the subsequent deficit. First, Finance Minister John Turner announced that oil companies could no longer deduct the royalties they paid from their taxable income. Companies would pay taxes on gross revenues, not net profits. "If the federal government had continued to allow corporations to deduct

Table 6.2   Average Annual Price of Oil, 1973–80 (C$/barrel)

Year	World Price	Canadian Price	Differential
1973	4.57	3.80	0.77
1974	9.74	5.83	3.91
1975	11.20	7.25	3.95
1976	11.46	8.53	2.93
1977	13.58	10.25	3.33
1978	14.76	12.25	2.51
1979	21.09	13.25	7.84
1980	35.75	15.58	20.17

Source: National Energy Board *Annual Report* for stated year.

provincial royalties," concluded economist Mandy Wahby, "it would have been equivalent to economic suicide. Its tax base had been severely eroded by the increases in royalty charges by the provinces, and at the same time, the federal government would have been allowing corporations to deduct operating and producing costs twice."[41]

Second, the government increased direct tax levies to 50 per cent and reduced the depletion allowance companies could claim from 33 per cent down to 25 per cent. This essentially reduced the percentage of exploration costs a company could deduct from its federal taxes.[42] The increased taxable income and higher levies diverted funds away from the Western Provinces to pay for purchases of offshore oil and served as a $3 billion a year subsidy. In part, an export tax on Alberta crude sold to the United States financed the deficit (Tables 6.3 and 6.4).[43] The main subsidy, however, came as price controls that fixed domestic prices of oil and gas at approximately half

Table 6.3   Canadian Crude Oil Production and Distribution, 1970–90, Mb/d (Thousands of barrels per day)

Year	Production	Imports	Exports
1970	1,382.1	568.9	669.8
1971	1,476.0	671.7	750.8
1972	1,698.4	769.6	951.3
1973	1,963.1	897.4	1,138.6
1974	1,844.1	797.0	910.9
1975	1,579.5	818.0	707.4
1976	1,444.8	755.3	465.1
1977	1,452.0	603.5	272.3
1978	1,434.1	544.5	174.3
1979	1,616.0	479.7	157.3
1980	1,541.1	438.7	93.5
1981	1,386.7	449.8	99.5
1982	1,371.5	275.9	156.2
1983	1,449.6	218.0	269.6
1984	1,534.0	215.5	313.7
1985	1,567.4	273.4	477.5
1986	1,564.7	356.6	589.7
1987	1,635.5	407.6	627.5
1988	1,717.4	451.7	710.6
1989	1,664.5	487.0	648.3
1990	1,659.4	543.0	656.5

Source: National Energy Board *Annual Report* for stated year.

Table 6.4   Canadian Crude Oil Export Taxes, 1973–80 (C$/barrel)

Year	Light and Medium Blend	Heavy Oil
1973	0.40	0.40
1974	5.19	4.10
1975	4.50	3.70
1976	3.74	2.90
1977	5.14	3.45
1978	5.39	3.75
1979	18.99	16.98
1980	28.70	16.33

Source: National Energy Board *Annual Report* for stated year.

world levels. Between 1973 and 1980, net federal revenues from oil and gas sources were probably negative amounts, since the government chose to peg domestic prices well below world levels while paying full market price for fuel imports.[44]

The new federal policy came as no surprise to the industry. Energy Minister Donald MacDonald met with John Poyen, President of the Canadian Petroleum Association, the day before announcing the price freeze and subsequent federal government policy. Poyen assured MacDonald that the industry would voluntarily accept a temporary crude price freeze, but was clearly uncomfortable when the Minister began discussing "windfall profits." Over the years, industry used the logical and basic argument that "profits" were essential to the health and future of the Canadian oil industry. The industry reinvested profits to offset finding costs, replacement costs, and development costs.[45]

Now Poyen sensed that the federal government was on the verge of taking policy in a new direction. "It appears that our Government economists are more and more ignoring those arguments and insisting that the revenues derived are subject to increasing scrutiny, the application of windfall profits disciplines and the imposition of greater economic rents, with little or no regard for the principles of reinvestment by the industry in future supply."[46] Poyen's hunch proved correct. While the federal government froze domestic prices, it allowed export prices to rise to world levels, an increase of approximately $0.40 per barrel. Instead of allowing the oil industry to retain the new profit, the federal government would take the entire increase as an export tax of $0.40 per barrel.

Meanwhile, the Province of Alberta and Syncrude finally hammered out a royalty arrangement in September 1973. Instead of pegging the crown royalty at a percentage of production value, as it had done with GCOS, the province agreed to take 50 per cent of Syncrude's net profits. The deal was contingent on the federal government treating the profit-sharing agreement as a normal tax royalty (i.e., that companies deduct their crown payments from their federal tax bill.) The federal government recognized the delicate nature of the agreement and granted Syncrude a special exemption from the federal tax policy. Donald Macdonald also committed the federal government to investigate the possibility that Syncrude could charge "adequate prices" for its product to make production viable. With the agreement in place, Syncrude began construction in the spring of 1974.[47]

OPEC dealt a staggering blow to world energy prices in October 1973 when the cartel invoked the oil weapon and immediately cut production 5 per cent. Cartel members agreed to keep cutting production in 5 per cent instalments each month the West ignored their demands.[48] OPEC's move surprised the federal government, and Ottawa could not escape the effects of the embargo. Eastern Canada was completely dependent on offshore oil, much of which arrived through U.S. pipelines. Thus, the oil shock was a mixed bag for Canada. On the one hand, Alberta's exports to the United States increased substantially, but on the other, larger Canadian imports from Saudi Arabia negated the benefits of increased exports.[49] Although multinational oil companies arranged those shipments, many in the Canadian government wondered if the oil companies would respect Canadian commitments when the United States was in need.[50]

Ottawa attempted to address the situation in December 1973 when Prime Minister Trudeau outlined his new energy policy. The Prime Minister asserted that his objective was "Canadian self-sufficiency in oil and oil products" before the end of the decade.[51] The federal government proposed a retrenchment of the Canadian oil industry. Rather than allowing the industry to operate along continental lines, Ottawa wanted the industry to operate along national lines. To hasten the change, Trudeau announced the creation of a national oil company, Petro-Canada, and indicated that the federal government would construct a pipeline to Montreal to relieve eastern Canada's dependence on imported fuel. "The creation of a national market for Canadian oil is one essential requirement of a new policy," said the Prime Minister.[52] To offset an expected shortage of three million barrels of oil in the winter of 1973/74, Ottawa slashed crude exports to the United States at least 10 per cent (exports of natural gas were granted an exemption). Combined with the provisions of the newly-created Foreign Investment Review Agency (FIRA) and the pre-existing Canadian Development Corporation (CDC), the federal government announced it would monitor American investment in the Canadian oil industry and begin to "Canadianize" the oil industry.[53]

Turning his attention to the oil sands, Trudeau believed that the chief obstacle preventing the effective development of the resource was the lack of research and development in the field. "Technologies must be developed which do not yet exist in order to permit the development of 85 per cent of the sands which are deeply buried." To help kick-start research projects, Ottawa pledged $40 million to oil sands research and promised that the proposed national energy company would be involved in Alberta's oil sands. In closing his remarks, Trudeau told Canadians that the uncomfortable reality was that the oil sands were a more expensive resource to develop than conventional oil. "The days of abundant cheap energy for Canadians must come to an end," declared the Prime Minister. Referring to Macdonald's September price freeze, Trudeau acknowledged that the federal government would allow the price of domestic oil to rise gradually toward the level needed to ensure profitable development of the oil sands "but not one bit higher."[54]

Before the federal government's energy announcement in December 1973, many in Alberta hoped that the oil sands would be able to capitalize on world events. Perhaps the crisis would cement the province's reputation as a reliable energy supplier to export markets. The new federal policy, combined with international trends, dampened industry enthusiasm. Canadian-U.S. relations clearly entered a new phase while the OPEC crisis prompted Ottawa to reconsider the continental orientation of the Canadian oil industry. The "special relationship" that characterized Canadian-U.S. relations since the end of the Second World War ended in 1971/72 and gave way to the "Third Option."[55] "Canadian public opinion," noted historian Robert Bothwell, "was going through a nationalist phase: the further left it was the more acute the concern it expressed about the United States, its government, its policies and its multinational corporations."[56]

As political economist John Erik Fossum later observed, the first oil shock provided Trudeau's Liberals the opportunity to develop a national energy strategy free from the pressures of U.S.-Canadian relationship. Sensing that the oil shocks were independent of the Cold War, many countries, including Canada, adopted policies that significantly diverged from those of Washington. The idea of a nationally oriented supply system, with Ottawa at its centre, "was based not only on the vision of a more independent Canadian stance vis-à-vis the United States but also on the assumption that OPEC would continue to dominate the market in a politicized world oil scene."[57]

Thus, the 1970s witnessed a renaissance of Canadian nationalism and a growing suspicion of American influence, in particular. In August 1971, national revenue minister Herb Grey presided over the compilation of a report that investigated foreign investment. Finding that Americans accounted for 77 per cent of all foreign investment in the country, the Grey Report issued its qualified conclusion that foreign investment had a moderately favourable effect on the Canadian economy overall, but that some problems did exist, most notably, the existence of "truncated firms." "Truncated firms" were those companies that could only provide a narrow range of activities and that were dependent on foreign technology or management. The Grey Report ultimately concluded that administrative intervention, on a case-by-case basis, of all new foreign investment could address some problems.

The federal government's decision to subject the oil industry to the provisions of the CDC and FIRA represented a slap in the faces of the multinational oil companies like Sun Oil who had invested much time, energy, and money in the oil sands. Consider that at the beginning of the oil shock in October 1973, Great Canadian Oil Sands – nurtured and funded by Philadelphia's Sun Oil – researched and developed forty-four of the sixty-nine technologies or processes related to the oil sands patented in Canada.[58] Where would oil sands development be if Ottawa had barred Sun Oil from investing in the sands a decade before? The Science Council of Canada pointed out that most of the knowledge and expertise in oil sands development technology resided with foreign-based companies. If the federal government intended to follow through on its decision to increase Canadian ownership of the oil sands, it would probably drive out the foreign-owned multinationals. Would the proposed advantages of increased Canadian participation in the oil industry offset the delay in oil sands development that would inevitably result?[59]

Trudeau's decision to set up export taxes and to slash oil exports to the United States also did not sit well with Albertans or their customers south of the border. In 1972, Canadian exports to the United States totalled 1 million barrels per day, much of which serviced the American Midwest and the New England states. The U.S. State Department argued that the Canadian decision would "not be well understood" in these areas, particularly if they endured a loss of heat or electricity during the winter.[60] "These Canadian actions violated the continentalist expectations that the government of Canada itself had helped stimulate not more than two years before," wrote Bothwell. "They may even have transgressed the 'special relationship.'"[61]

Washington's complaints resonated in Alberta. It did not take long for both the province and industry to add to the list of grievances. The decision to freeze domestic oil prices – even if temporary – hurt the petroleum industry as a whole, but was a particular hardship for Great Canadian Oil Sands and Syncrude.[62] The federal government's decision to prevent oil companies from deducting their provincial royalties cost GCOS $3.6 million in 1974. The inflationary environment and lack of accord between the federal and provincial governments created "considerable doubt" about the future of oil sands development. "Until these problems are resolved then it would appear that the Athabasca oil sands will play only a minor role in providing Canada's future energy needs," warned the GCOS Annual Report.[63] Meanwhile, inflationary pressures pushed the capital costs of the Syncrude project from $650 million at the end of 1972 to $1 billion in July 1973.[64] The consortium's investors could not recoup their initial investments if the federal government froze the price of its product.

The most disturbing aspect of the federal energy plan was the growing conflict between Alberta and Ottawa. The provincial premier called the federal energy action "the most discriminatory action taken by a federal government against a particular province in the entire history of confederation." Lougheed then introduced new legislation to retaliate against the federal measures. His move displayed the province's control over production and pricing of crude oil.[65] From their ringside seats, the industry saw that neither side showed concern for the overall health of the oil industry. Both levels of government were more intent on capturing larger shares of the "windfall profits" and asserting prerogatives in the fields of royalties and taxation. Nevertheless, research continued on the oil sands mostly because developers believed that both levels of government would have to recognize that production from the sands was necessary. Self-interest alone dictated that policies would evolve that would make oil sands development profitable for all parties. If the federal-provincial conflict continued, the lack of long-term – or even consistent – guidelines would inhibit the pace of development.

In retrospect, the federal energy policy prevented the differences of opinion between the oil industry and the provincial government from emerging publicly. Indeed, during his brief term in office, Lougheed had already substantially raised the maximum royalty level from 16.7 per cent to 23 per cent. A few months later, he unilaterally declared that royalty rates would rise with the world price of oil. This decision was made without consulting the industry beforehand and it angered company executives who grumbled that the province had broken a contract. Doern and Toner point out that "this was virtually the first indication the industry had that the government of Alberta perceived that it had a set of interests related to oil and gas that were distinct from those of the industry."[66]

Lougheed mistrusted "big oil," and the industry, in turn, believed that the province misunderstood them. At least the two sides could bond over their mutual enmity for Pierre Trudeau – instead of dividing his opponents, Trudeau drove the industry and province into the arms of one another. Ottawa's decision to create Petro-Canada was a particularly egregious affront to the oil industry. Without the creation of the Crown corporation they might have been willing to go along, albeit grudgingly, with the other provisions of the federal plan. But in Calgary's boardrooms, the creation of a state-owned oil company implied that Ottawa did not, and could not, trust the industry.[67] While the industry did not always agree with Lougheed's government, the province's policies remained less objectionable than those of the prime minister.

RETAINING "FLEXIBILITY"

With the second commercial oil sands plant now underway, government and industry attention turned to creating the necessary conditions for a third plant. Despite the industry's desire that the province adopt a comprehensive oil sands strategy, the province steadfastly refused to commit itself to a long-term policy. Nevertheless, both the province and industry associations continued to meet and discuss ways of improving oil sands development. While the Province of Alberta was not interested in articulating an overarching strategy, it prepared to implement a series of temporary measures to guide further development.

In the opinion of the Alberta Department of Federal and Intergovernmental Affairs, none of the bottlenecks holding up oil sands development related to government policy. The province focused on labour and supply shortages and the slow pace of developing appropriate *in situ* production methods.[68] Premier Lougheed clarified the importance the province attached to the final item a few weeks later when he announced the creation of the "Energy Breakthrough Project." Capitalized at $100 million, the province launched an all-out effort to develop new technology to mine the oil sands. It also created the Alberta Oil Sands Technology and Research Authority (AOSTRA) to guide future research and development.[69]

While the industry generally liked the provincial initiative, CPA briefing papers and memoranda suggest that the Association remained more concerned with indirect measures available to government to influence development. For synthetic producers, taxation, royalties, pricing and export policies were the crucial questions in all negotiations regarding oil sands development. Close examination of the continuing negotiations reveals that government and industry had different perspectives on the risks involved with oil sands development. The disparity between the two positions frustrated the CPA. Eventually, the organization perceived the need to prepare and distribute a report detailing the risks associated with oil sands production.[70]

Ottawa shared Edmonton's assessment about the risks involved with oil sand ventures. Federal Energy Mines and Resources staff maintained that the oil sands were "significantly lower risk" ventures than conventional crude operations. Oil sands producers knew *exactly* where the oil sands were and simply had to develop them. Thus, the federal government asserted those oil sands projects did not require revenue from conventional production to help finance operations. New oil sands developers would secure high debt financing more easily from their creditors than conventional producers. The CPA sharply disagreed. Current federal policies were inconsistent in several crucial areas – energy self-sufficiency on the one hand and limiting foreign investment on the other.

Overall, the policy contained several disincentives for oil sands production. The current federal and provincial tax regimes did not give companies enough return on their investment to make exploitation of the sands attractive. If, for example, demand for crude oil dropped sharply, the tight economics of oil sands development would force the producing company to absorb huge losses. No responsible corporate officer could commit his company to such a project, particularly in marketing a volatile commodity like oil. However, guaranteeing oil sands producers a share of the market would make finding potential investors much easier.[71]

The question presented to the provincial and federal governments was quite simple: could government assure oil sands producers a market for their production? Concerns lingered about consumer willingness to pay higher prices for synthetic crude if cheaper imports became available. Could excess production from the oil sands get an exemption from Trudeau's policy eliminating Canadian crude exports to the United States? An affirmative answer would assure the industry that there was an acceptable risk in continuing to devote time and resources to the oil sands.

Government and oil producers continued to talk past one another until the petroleum association played the shale-oil card. Fed up with the endless rounds of discussions, producers warned that Canada's oil sands policy stood in direct contrast to recent developments in the American government's oil shale policy. Corporate executives hinted that the industry might be induced to greener pastures south of the border. A few months earlier, on November 7, 1973, President Nixon announced the creation of Project Independence, an ambitious synthetic fuels plan designed to wean the United States from foreign oil by 1980. Invoking such scientific triumphs as the Manhattan Project and the Apollo space program, Nixon believed that a similar drive by industry and government on oil shale would end triumphantly. "It was clear that when the American people decided a particular goal was worth reaching," recalled Nixon in his memoirs, "they could surmount every obstacle to achieve it."[72]

Although the CPA acknowledged that the two sources would not compete directly for the same market share, perhaps there was an opportunity in the fact that both countries sought to develop synthetic crude simultaneously. More to the point, both countries were competing for capital investment from the same corporations.[73] As the CPA reminded the province on May 9, "we must expect capital to be attracted first to the development which provides the most incentives." Oil shale boasted better lease terms, as well as more friendly rent and royalty payments. The implication was clear. If the province did not modify its stance, development of the oil sands would lag behind that of oil shale.[74]

The CPA's reference to the oil shales did little to help their case. Conventional wisdom dictated that the oil sands enjoyed two distinct advantages over every other synthetic fuels project being funded, or about to be funded, in North America. Alberta remained several years ahead of any other nation in knowing how to transform oil sand into oil. Some analysts estimated that even if the Americans launched some crash synthetic fuels projects, Alberta would already be using second-generation technology to surpass their output. Alberta also enjoyed the luxury of having less expensive, and less polluting, processes than those involved with turning coal and oil shale into liquefied fuel.

In stark contrast to the speed with which the second generation of oil sand projects proceeded, Project Independence struggled to get off the ground. Indeed, Nixon's televised national address on November 7, 1973 is remembered more for his pledge to ignore mounting demands for his resignation than it is for the details of his energy package. Announced as it was at the height of the Watergate scandal, perhaps it was inevitable that Project Independence encountered stiff congressional and business resistance before being relegated to the dustbin of history. Nevertheless, economic historian Diane Kunz later concluded that one of the major flaws of Project Independence was that its goals were clear – a massive effort to develop a synthetic fuels program – even if the means of achieving it were not. "A key issue was whether government fiat or free market action should lead the way." [75] The question was never resolved, and by the end of the decade, costs of production for the oil sands dipped to approximately $20 per barrel, while costs on shale oil were estimated at approximately $25 a barrel, and production costs for products resulting from coal liquefaction were at least $40 a barrel. [76]

Having failed after invoking the oil shale card, a CPA memorandum in July 1974 suggests that the organization modify its tone but retain the main thrust of its message. Specific references to American oil shale policy were dropped from the draft of the CPA's submission to the government circulated to members, but the CPA reiterated that a combination of government policy, economic attractiveness, and technological advancement would determine the pace of future development. The industry would be prepared to move forward "*provided a framework is set by government which will act as the stimulus for a healthy business climate in which industry can operate with a reasonable degree of predictability.*" [77] Industry could assess and accept the risks and reach investment decisions in instances where the market operated freely, but the reality now was that oil sand projects were "critically dependent on future federal and provincial government action and legislative and regulatory stability." Again, the CPA reminded the province of the need for a comprehensive policy. [78] Despite the CPA's best attempts, the Lougheed government clearly preferred to trade barbs with the ruling federal party on most energy-related issues.

Verbal sparring between Ottawa and Edmonton over oil policy continued until December when developments in the synthetic industry forced the two sides to set aside their differences on another front. On December 4, 1974 Atlantic Richfield announced that it was pulling out of the Syncrude project because the company's heavy investments in Prudhoe Bay precluded support for the Athabasca project. It now fell to the remaining partners, Imperial Oil, Cities Service and Gulf Canada to find another investor willing to assume a 30 per cent stake in the project. Bill Dickie, Alberta's Energy Minister, contacted the federal government and invited the provincial governments of Ontario and Quebec to participate in the project. Although the Province of Quebec declined, both the federal government and the Province of Ontario indicated their willingness to join in the venture. On February 3, 1975, the consortium's new composition emerged after a meeting in a Winnipeg hotel room. The partners in Syncrude were now Imperial Oil (31.25%), Cities Service (22%), Gulf (16.75%), the federal government (15%), Alberta (10%), and Ontario (5%).[79]

The revised Syncrude deal had as many opponents as it did supporters. Critics on the left charged that the industry and the Province of Alberta got a "sweetheart deal" from the federal government. Syncrude gained an exemption on the recent federal decision to disallow provincial royalty payments to be deductible from federal taxes. On pricing, Ottawa allowed production from Syncrude to move closer toward world prices, provided such prices were "fair." Syncrude's production would also be exempted from any pro-rationing scheme, guaranteeing that they could market the plant's full production, even, if necessary, in export markets.[80] Aside from the actual terms of the deal, the Syncrude negotiations "provided a much-needed opportunity for the two levels of government to prove they could work together."[81] Ottawa yielded to the industry power play, wrote Doern and Toner, and chose to display its willingness to enhance "Canadian oil security by strengthening Canada's domestic supply of accessible oil."[82]

PROTECTING THE ENVIRONMENT
IN THE AFTERMATH OF THE EMBARGO

Environmental protection provided a notable – but nonetheless temporary – avenue for inter-government co-operation. Construction of the first GCOS plant encountered few environmental restrictions specifically designed to monitor for oil sands development. In fact, by the time Syncrude prepared to move forward with the construction of the second oil sands plant, the province did not even possess the most basic "baseline" data about the environment of the Athabasca area – Intercontinental Engineering completed the only major survey of the environmental impacts of oil sands development in 1973.[83]

Belatedly, both levels of government moved to rectify the situation after the size and scale of the proposed Syncrude plant became clear. Late in 1973, both Alberta Environment and Environment Canada separately produced internal reports recommending the creation of a comprehensive environmental research program. Each suggested the creation of a research program lasting for ten years (renewable after the first five years) with total costs pegged between $30 to 40 million. Two years later, in February 1975, Edmonton and Ottawa signed the Canada-Alberta Agreement creating the Alberta Oil Sands Environmental Research Program (AOSERP). With a first-year budget of $4.5 million ($2.5 million from Edmonton, $2 million from Ottawa) AOSERP's mandate specified that the project would investigate the environmental impact of oil sands development on the air, land, water, and people of Northern Alberta. The province accepted administrative responsibility for the program but a two-person steering committee (one representative from Alberta Environment, the other from Environment Canada) determined policy direction.[84]

AOSERP established eight separate research committees to conduct investigations into a variety of different areas. The program's ambitious first-year operations plan identified six objectives for the program:

1. To receive approval for research projects
2. To establish a program coordination and administrative capability
3. To design in detail ongoing research projects
4. To assist project mobilization where required by detailed planning, recruiting and staffing, and provision of resources

5. To establish some level of field support for all projects
6. To satisfy the federal/provincial Agreement regarding reporting.[85]

The steering committee intended to use the first two to three years of the program to determine the "baseline" environmental conditions directly and indirectly affected by oil sands development, despite the fact that GCOS had begun operations nearly a decade earlier. "Decisions to investigate 'baseline' environmental conditions," concluded AOSERP's interim report, "therefore, had to take into account that in some respects the area already had been impacted by industrial development."[86] Indeed, when AOSERP's mandate began in 1975, it would be measuring the impact of two previous oil spills from the GCOS plant in Athabasca River in addition to the cumulative environmental impact of a large surface mining operation seven years removed from the beginning of commercial operations in 1968. Clearly, the proposed "baseline" testing would establish environmental standards of the Athabasca region with one oil sands plant operating and the construction of a second underway.

Although the policy trade-offs between the environment and energy development might be disturbing to contemporary audiences, it is important to remember that the majority of agencies responsible for protecting Canada's environment were not founded until the 1970s. The federal Department of Environment (the precursor to Environment Canada), for example, was established in 1971. Environmental Impact Assessments (EIA) required for proposed development projects were not required business practice until the creation of the Environmental Assessment and Review Program in 1973.

The reality is that prior to 1975, neither Edmonton nor Ottawa articulated a coherent environmental strategy to guide development of the oil sands. In fact, the underlying concern for both federal and provincial governments was not protection of the environment. Rather, elected officials needed to determine the best way of producing enough oil for Canadians to heat their homes, light their streets, drive their cars, and maintain their standard of living in the shortest time possible. Furthermore, the spectre of another OPEC embargo loomed behind many policy choices in the 1970s. So, too, did fear. Fear of the long Canadian winter. Fear of gas shortages. Fear of angry, and cold, voters. Fear, in short, of living out the slogan emblazoned on the bumper sticker adorning a number of Alberta vehicles to: "Let the Eastern Bastards Freeze in the Dark!"

Despite the emergence of environmental concerns into the policymaking process, it is a testament to the depth of the perceived energy crisis that no provisions were made to halt construction of an oil sands plant if ecological problems were identified. On April 17, 1974 Provincial Environment Minister W.J. Yurko addressed the Engineering Institute of Canada Conference in Edmonton. Although his remarks emphasized the need for government to collect and assess "baseline" data on "surface and ground water hydrology, soil characteristics, biological resources, flora and fauna, sub-surface and [surface] geology, and climatology and meteorology characteristics," development would proceed rapidly thereafter. "Once the momentum of building oil sands plants is established," said Yurko, "it will be sustained." Alberta's environment minister doubted it would be politically feasible for any government to hold back oil sands development once a trained workforce of tradesmen, technicians, engineers, and scientists existed. Critically, Yurko concluded that "the market demand for the resource is not expected to subside." [87]

TOWARDS A THIRD PLANT?

For a brief moment in the aftermath of the Syncrude negotiations and the federal-provincial co-operation on environmental issues, it appeared that Edmonton and Ottawa had reached some sort of accommodation. Both sides appeared willing to sit with industry to decide what incentives were necessary to kick-start construction of a third oil sand plant. In early 1977, the two levels of government established a Joint Task Force to review the range of economic conditions (capitalization, plant size, investment opportunity, royalties, and taxation) needed to begin construction on a third oil sands plant.

The perceived thaw in federal/provincial relations encouraged the CPA. Hopeful that these developments signalled the willingness of government to join with industry, the CPA forwarded an informal brief to the province on February 24. The Association believed that reducing the royalties alone would not be sufficient to encourage future oil sands development. Instead, both levels of government would have to recognize that income tax relief and "reasonable" royalties were essential to the industry's future. The scarcity of available capital to finance industrial projects, and the federal government's Canadianization plans, meant that the public sector would have to aid development.[88]

The CPA elaborated its position a few weeks later when it filed a formal submission to the Joint Task Force. Industry's message was difficult for government to swallow. The cost of oil sands development under current tax regulations led the Association to conclude that government should expect no significant royalty. Rising material and labour costs, due to inflation, would consume profits, even if producers received 1976 world oil prices of $13.86 per barrel. Nevertheless, the CPA suggested a number of ways the Task Force could enhance the attractiveness of oil sands projects.

1. Establish a guaranteed floor price for synthetic producers
2. Guarantee the debt incurred by any new commercial venture
3. Enable plant owners to have "first call" on their production
4. Grant synthetic fuels access to export markets
5. Reduce or defer royalty payouts until investors recouped their initial outlay [89]

When the CPA met with representatives of the province and the federal government in May, government response was disappointing. The federal government recognized that the construction of a third oil sand plant should be encouraged, but not at any cost. The CPA's terms were clearly beyond anything Ottawa considered reasonable. There was general consensus between the parties that construction on a third plant would not be viable without substantial alterations to the fiscal regime but the meeting concluded without agreement.[90]

Much to the industry's dismay, however, the final report's main conclusions simply echoed the arguments that the Department of Energy Mines and Resources had put forward in April 1974. From the private sector's point of view, both levels of government were merely paying lip service to the importance of oil sands development. Where government saw certainty, corporate officers saw nothing but risk. As a whole, the industry believed government completely divorced from reality the recommendations contained in the report; government parameters were woefully inadequate, particularly concerning royalty and taxation issues. The oil patch's reply was clear and united: without an improvement in the province and the federal government's commercial terms, they could not foresee a time when industry would begin development of a third oil sands plant on its own.[91] The Canadian Petroleum Association temporarily disbanded its Oil Sands and Heavy Oils Committee at the end of 1977 after sensing that developments in oil sands policy were unlikely.

Perhaps more ominously, the chill between Edmonton and Ottawa returned in September 1978 as the federal government backed out of an agreement to raise Canadian oil prices to within $1 per barrel of the world price. Although it may only have been a coincidence, the Liberal government in Ottawa also unilaterally announced at roughly the same time that it would withdraw federal funds from AOSERP effective March 31, 1979. Understandably disappointed by the loss of federal monies, the provincial environment minister was more disturbed by the fact that Alberta alone would be forced to bear all the social and environmental costs of developing a natural resource for the "national interest."[92] Nonetheless, Edmonton quickly pledged to continue funding for the program until the end of 1980.[93]

After the Syncrude negotiations, the time was ripe for both the government and industry to consider additional joint ventures to develop the oil sands. Instead of striking while the iron was hot, the momentum generated by Syncrude gradually waned. Despite a growing Canadian dependence on offshore supplies, there was no progress on the construction of a third oil sands facility. Public criticism suggested that the lack of positive government policies hampered the oil sands' potential. The federal government admitted that it faced increased pressure to compete with Jimmy Carter's programs south of the border.[94] Without the construction of additional commercial oil sands projects, the public began to wonder how Trudeau would achieve his goal of national energy self-sufficiency.

CHAPTER 7

Lost Decade: The National Energy Program and the Collapse of World Crude Prices

Well, welcome to the eighties. – *Pierre Elliott Trudeau,*
Campaign victory speech, February 18, 1980

Many conventional producers decided to leave Alberta during the height of the energy feuds between Edmonton and Ottawa. Seeking stability and a tranquil investment climate, they turned to the "frontiers" of oil and gas development – primarily the Arctic and the Hibernia project in offshore Newfoundland. However, the pattern changed with the Iranian Revolution in 1978. The price of oil more than doubled, leaping from US$14 to US$34 a barrel. Overnight, perceptions about the investment climate in Alberta changed and many companies scrambled to make their mark while energy prices remained high. If the industry hoped that the simmering feud between Peter Lougheed and Pierre Trudeau over energy policy had cooled in the interim, they were sorely disappointed. The spike in world oil prices, resulting from the turmoil in Iran, ignited the oil policy fuse again. Soon enough, the unresolved conflict between the federal and provincial governments would push towards a final solution.

Nonetheless, a sense of unbounded optimism permeated the oil industry in the late 1970s as it experienced a new burst of growth. In 1974, approximately three hundred oil companies operated in Calgary; five years later, the total exceeded seven hundred. Many newcomers were small firms, dependent on bank financing. "Debt funding was the norm for these smaller firms," wrote political scientists Bruce Doern and Glen Toner. "Psychologically and ideologically, there was a contagious and invigorating spirit of free enterprise."[1] Many seemed to think the boom would never end. "The opportunities were boundless," wrote journalist and energy author Peter Foster. "If you were hardworking, or smart, or lucky, then you just had to make it. If you were all three, millionairedom was almost inevitable."[2]

If the trend in the conventional oil sector was to smaller independents, the exact opposite occurred in oil sands and synthetic fuel projects. The drift to "bigness" that began with Syncrude in 1978 became a veritable stampede as the major Canadian companies – Shell, Imperial, and Gulf – made commitments to larger oil sands projects.[3]

ALSANDS AND IMPERIAL COLD LAKE

Originally excluded from oil sands development in 1962 by the 5 per cent provision in Alberta's Oil Sands Development Policy, Shell put its oil sands projects on the backburner until more favourable conditions emerged. Less than a decade later, Shell Canada was prepared to move forward. The company altered its development strategy slightly and initiated a large oil sands project based on the same mining techniques used at GCOS and Syncrude.[4] After careful study, Shell concluded that the proposed project, which involved constructing a 137,000-barrel per day facility over nine years, presented substantial financial risks for a single company. Company estimates pegged start-up costs at approximately $5.1 billion. Shell considered shouldering the cost alone, provided the federal government granted the company price assurances on production to guarantee a minimum payout. When the government of Canada turned down Shell's appeal in 1977, the company looked for partners to share the financial exposure and risk. Finally, in August 1978, a participation agreement between Amoco Canada (10%), Chevron Standard (8%), Dome Petroleum (4%), Gulf Canada (8%), Hudson's Bay Oil and Gas (8%), Petro-Canada (9%), Petrofina Canada (8%), Shell Canada (25%), and Shell Explorer Limited (20%) ensured the project would go forward.[5]

Table 7.1   Initial Investment Required to Produce a Useable
Barrel of Oil from the Oil Sands, 1968–88

Project	Year	Investment Capital Required per Daily Barrel
Suncor (GCOS)	1968	$7,800
Syncrude	1978	$20,000
Imperial Cold Lake*	1987	$42,000–45,000
Alsands*	1988	$35,000–40,000

*Denotes projected completion date and 1979 estimated project costs.

Source: John E. Feick, "Prospects for the Development of Minable Oil Sands,"
Canadian Public Policy IX (1983), p. 299; A. Janisch, "Oil Sands and Heavy
Oil: Can they Ease the Energy Shortage?" *The United Nations Institute for
Training and Research: The Future of Heavy Crude Oils and Tar Sands.* First
International Conference, 1979, p. 40.

Despite the years of research that had gone into exploring alternative
means of producing oil from the sands, companies still used the combination of strip mining and Karl Clark's hot-water separation process for their
operations. Using older separation methods, however, did not limit costs.
Oil sands operations were huge and required thousands of workers to build
and maintain them; capital costs increased substantially with each successive project from GCOS onward. The investment capital required per daily
barrel produced by successive facilities more than doubled each decade,
reflecting the inflation of the late 1970s (Table 7.1).[6]

Given the enormous construction, production and maintenance costs
involved, companies chose not to experiment with unproven technologies when they could rely on proven methods. The Alsands Project Group
(APG) bluntly told others within the industry "we can't afford to get involved in a $5 billion research project."[7]

Imperial Oil also made plans in 1979 to move forward with its projected $6 billion, 140,000-barrel per day, Cold Lake Project to the south of the Suncor and Syncrude plants.[8] As the first *in situ* project, the Cold Lake initiative was important for both the industry and the provincial government. If successful, it would unlock the remaining 90 per cent of the oil sands reserves that producers could not exploit through conventional mining techniques. Imperial proposed to drill approximately 980 wells at the start of the project and then add 350 wells per year after that until the total number of wells reached 8,300.[9] Once the wells were in place, Imperial would inject high-pressure steam into the deposit for one month. The field would rest for a month to allow the moisture to soak into the deposit that would enable the wells to produce for three to six months.[10]

The CPA reactivated the Oil Sands and Heavy Oils Committee in December 1979. With two new "mega-projects" underway, representing approximately $11.1 billion in investment, and forty-five experimental projects operating throughout the Athabasca area, the time seemed appropriate.[11] In early January 1980, the committee learned that the province intended to hold discussions with the oil sands industry regarding future work for leaseholders. Once again, the industry believed that the province intended to introduce a new policy that would take into consideration the special needs of oil sands developers. This new policy would, hopefully, streamline the application process and inaugurate a new era of intergovernmental and industry co-operation.[12]

In June 1979, F.H. Allen, President of Amoco Canada and Chair of the CPA's Board of Governors, addressed many of these concerns in a speech to a United Nations conference. In that forum, Allen remarked that developers were growing weary of the constant bickering between Edmonton and Ottawa. Instead of improving over time, relations continued deteriorating. Company executives now spent countless hours on planes between the two capitals, engaged in a form of shuttle diplomacy to force a solution. "We have grown accustomed to first talking with one level of government and then shuttling for a series of discussions with the other level, and then going back to the first," noted Allen. Nevertheless, more than one industry observer concluded that the overlapping of jurisdictional authority and lack of coordination had seriously delayed development of Canada's oil sands and heavy oil projects. Pointing to reports that indicated demand for oil continued to rise while conventional production declined, Allen argued, "We no longer have time to hesitate." All parties needed to set aside their differences in order to find a solution.[13]

PRESIDENT CARTER'S ENERGY SECURITY CORPORATION

Developments in the United States also heightened the Canadian industry's sense of urgency to implement a new policy. Politicians on both sides of the border needed to do *something* in the face of the energy crisis and embraced synthetic fuels projects as one potential solution. Oil shale, the oil sands' traditional rival, seemed poised to make a major breakthrough when placed at the centre of a new, and ambitious, synthetic fuels initiative. In 1979, the Carter administration began putting together plans to create the massive state-sponsored Energy Security Corporation to manufacture synthetic fuel alternatives to offset the loss of OPEC's conventional crude and the hoarding of supplies from alternative sources, including Canada.

Despite the best efforts of the president to implement a national energy policy in 1977, the stakes increased when the Shah's regime collapsed in 1979. Fear gripped some members of the Carter administration as the revolution removed the two million barrels of OPEC oil from the world market.[14] Before the end of the year, Eliot Cutler, Carter's chief energy adviser, glumly noted, "We have no assurance that world oil production will ever exceed sixty-five million barrels. We are living at the knife edge – dangerously close to the margin between energy supply and demand."[15]

John O'Leary, head of the Federal Energy Agency and the Deputy Secretary of Energy, and F. David Freeman, argued forcefully that they should set up a comprehensive synthetics program to offset supply shortfalls. After a briefing by the CIA and the State Department in late May, O'Leary said, "We have the edge of disaster peeking out at us."[16] The faith of politicians in large synthetic fuel projects to meet consumer demand was contagious. If the price of gasoline rose by ten cents per gallon, *Chemical Week* estimated the revenue could produce one million barrels of synthetic fuels per day. "Such a program could produce 5 million barrels per day of capacity in synthetic fuels in 1990 and put the country on the road to adequate supplies of energy."[17]

Against the backdrop of growing supply shortages and increasing attacks on his leadership, President Carter unveiled his second energy package on July 15, 1979. Stating that the United States faced a "crisis of confidence," the president announced that his administration would begin a synthetic fuels program. The administration would allocate $88 billion to produce 2.5 million barrels per day of synthetic crude from coal and shale oil. "We've got to stop crying and start sweating, stop talking and start walking, stop cursing and start praying," argued Carter. "The strength we need will not come from the White House, but from every house in America."[18]

From this point of view, perhaps the real crisis created by the collapse of Iran was the effect it had on the public perception of energy issues and proposed solutions. In December 1978 when Iran stopped exporting oil, other countries took steps to offset the loss of Iran's daily production of six million barrels. Saudi Arabia increased its production by two million barrels per day and other OPEC countries followed suit. Thus, the amount of oil produced in the first quarter of 1979 was only two million barrels per day lower than the last quarter of 1978.[19] "It was not the magnitude of loss of supplies from Iran that created the second oil crisis," argued political scientist Mohammed Ahrari. Rather, the second oil crisis was the product of the combination of distribution bottlenecks and the onset of a siege mentality in consuming nations.[20]

Carter's synthetic fuel project dwarfed anything the Canadian government could dream of attempting and illustrates a fundamental difference between synthetic fuels policy in the Canada and the United States. Canada needed production from the oil sands to sustain its oil industry and focused on the development of alternative fuel sources as a way to become energy self-sufficient. But some in the United States government regarded synthetic fuel projects as the means to break OPEC's stranglehold on prices. Prices on the spot market skyrocketed as competition between the industrialized nations increased to make additions to their fuel stocks. In early 1979, OPEC producers began adding a monthly premium to their official price. By March 1979, spot prices rose by 30 per cent and refined products by as much as 60 per cent. "Exporters," noted oil historian Daniel Yergin, "were abandoning any notion of an official price structure. They would charge whatever the market would bear." [21]

In May 1979, Arthur Okun, a Harvard economist working in the U.S. Treasury Department, implored the administration to begin its synthetic fuels project. The effort alone would create a sense of uncertainty among world crude suppliers. "Everyone is predicting into the indefinite future continuing increases in the price of petroleum," wrote Okun. "Hence, oil in the ground is universally viewed as a great investment." How could the United States break the premise that oil in the ground is "a great, riskless" investment? "That requires the realistic prospect of a genuine technological breakthrough within this century in one or more directions." [22]

> The United States – appropriately in combination with the other major industrial oil consuming nations – must make the commitment to the big technological drive. We must set the proper priorities, and we can put the risk back into oil in the ground. That long-term venture would have an immediate payoff by changing the restrictive supply policies of the OPEC nations. Shaking confidence in the oil-in-the-ground as a great investment provides the only happy ending that I can see to the present energy crisis. [23]

Clearly, the Trudeau government did not entertain similar thoughts about energy policies. As the Carter administration wrestled with the logistics of implementing the Energy Security Corporation, Progressive Conservative leader Joe Clark narrowly defeated Trudeau's Liberal Party in a hastily-called election. Clark immediately signalled his willingness to change Canadian energy policies – including dismantling Petro-Canada. [24] But when voters unexpectedly returned the Liberals to power in the spring of 1980, Clark's plans came to an abrupt end. [25]

TRUDEAU, LALONDE, AND THE NEP

After a brief hiatus, happenstance and poor counting by Clark's Conservative government returned the Liberals to power. Given this unexpected chance to redeem themselves after an ignominious departure months before, Trudeau and his chief lieutenants became obsessed with leaving their mark on history.[26] One way to do this would be to focus the energies of the state on a few important issues – the constitution, the economy, and energy. Trudeau appointed Marc Lalonde, a combative and fiercely loyal member of the Cabinet, as his Energy Minister. Lalonde promptly began work on the government's energy policy as part of a two-pronged effort to curb the powers of the provinces and to reassert Ottawa's primacy. First, Trudeau would confront the provinces as a whole in public debates on the constitution to display the power of the federal government in no uncertain terms. Second, and most important, Trudeau would attack and neutralize the political power centres of Alberta and Quebec, the two most formidable provincial opponents of increased federalism.[27]

Increasingly, Trudeau believed that the solution to all of his problems with the provinces lay in unilateral action. Ottawa must act decisively to arrest the devolution of power away from the central government. He thought more federalism, rather than less, would produce desired results and would tame the growing power of the provinces. With a clear mandate from the electorate, Trudeau's Liberals could govern for at least five years before having to defend their record in an election. Trudeau thought the time was ripe to carry out radical economic reforms. "From 1976 on," recalled Trudeau, "my real battle was to prepare Canadians – and especially Quebecers – for the day when the federal government might have to break convention and make an appeal to London without having gained unanimous provincial consent, and perhaps unilaterally." [28]

Trudeau's unilateral approach on the energy question and Canada's industrial policies were designed to complement the constitutional battle – there had to be some advantage to entice Canadians to support his plan. In their definitive biography of Pierre Trudeau, Stephen Clarkson and Christina McCall concluded that the interventionist approach adopted by the Liberals committed the Canadian government to "a philosophy of governance already in deep trouble around the world, as democratic administrations encountered increasing difficulty in funding social programs and sheltering their entrepreneurs and workers from the competitive pressures of rival economies." [29] But, given the large Liberal majority in the House of Commons, Trudeau could ram his agenda through and impose his will on the oil industry with impunity. "That was the spirit after 1980," wrote Trudeau. "We had demonstrated that a ponderous, careful pace doesn't get you more co-operation; perhaps it will get you less. So we just set out to get some things done." [30]

The federal-provincial dispute had not mellowed during Trudeau's brief absence from power during 1979 to 1980. If anything, the acrimony between Edmonton and Ottawa intensified, much to the disappointment of many Canadians. In 1979, Joe Clark successfully campaigned on a platform of improving federal-provincial relations and wasted little opportunity in extending olive branches to various provincial capitals. The Liberals were far from amused. From their seats on the opposition side of the House of Commons, Trudeau seethed that Clark reduced the role of Prime Minister to little more than "head waiter to the provinces," while Jean Chrétien called Clark the "Neville Chamberlain of Canada." [31]

Despite Clark's vision of Canada as a "community of communities," complete with a decentralized power structure, Peter Lougheed continued to approach energy negotiations as a zero-sum game. From Lougheed's point of view, it did not matter that Clark was Prime Minister - the entrenched bureaucrats at Energy, Mines, and Resources still formulated policy. The Clark government's stance on energy issues hardly differed from those of Trudeau because the same people authored both. But in politics, where perception is king, the public saw the negotiations quite differently. Lougheed was not jousting with Pierre Trudeau. In Trudeau's place stood an affable, if unspectacular, politician and fellow Albertan, no less. Rather than a clash of titans, the battles seemed vaguely lopsided; the burly Lougheed versus the overmatched Clark.

Although Alberta's negotiating stance did not change noticeably with Clark in office, journalists chastised Lougheed for his "mean-spirited" and "greedy" treatment of the new Prime Minister. [32] Public perception greatly aided Trudeau's Liberals in the 1980 election. "The energy policy that Trudeau announced during the 1980 election was designed as a clear signal," recalled Liberal pollster Martin Goldfarb. "We knew where we were going ... we knew what was best for Canada, that we will speak for the little guy." [33] If Edmonton could not reach an agreement with the most sympathetic federal government of the past decade, perhaps it was time to settle the energy question unilaterally.

In the meantime, the increase in world petroleum prices during 1978–79 made Alberta oil sand investments more attractive. Synthetic producers constantly lobbied the federal government to allow them to charge world prices for their output, instead of the government-imposed "made-in-Canada" price far below world levels. In Joe Clark's brief reign, Ottawa allowed the price for synthetic crude to rise to the world price. The price of conventional crude reached only 75 per cent of international prices. "Taking $20 US as the prevailing world price by mid-1979," pointed out energy writer Peter Foster, "Syncrude's participants were looking at additional gross profits of more than $1 million a day at full capacity."[34]

Upon returning to power in 1980, Trudeau attempted to reverse this decision. Ottawa announced that it would invoke *force majeur* on the price schedule for the Syncrude plant. Once in place, this meant the company could only charge a "made-in-Canada" price instead of the full world price. This was hardly realistic, since the cost of synthetic oil production was running closer to the world price rather than to the artificially low domestic price. Because of several technical problems with the new facility, the cost of production averaged $30 per barrel in 1979. Gulf Canada and Imperial balked at the government's unilateral decision. The companies argued that production delays could force the plant to remain on the backburner for a year. Energy minister Lalonde was not amused and claimed the oil sands industry was blackmailing Canada.[35]

However, the second oil shock also exposed fundamental weaknesses in Trudeau's earlier energy plan. As world crude prices increased, the federal government could no longer afford to subsidize the consumption of imported fuels for Eastern Canadians. Within Canada's borders, the struggle between energy "haves" and "have-nots" played out in an increasingly acrimonious atmosphere. "It was a whole new situation," wrote Trudeau in his memoirs. "One that Canada couldn't possibly handle under the existing agreements that were giving the federal government only a 9 per cent share of oil and gas revenues, while the producing provinces got 50.5 per cent and the industry 40.5 per cent."[36]

Despite repeated, if perfunctory, attempts to agree on petroleum pricing, no agreement was forthcoming when the old package lapsed on July 1, 1980. World crude prices jumped again in 1979 and 1980, reducing the wellhead price of Canadian oil to less than half the world price.[37] Negotiations between the province and the federal government on a new energy pact had broken down nearly a month before. Both sides believed they enjoyed strong negotiating positions. The federal government claimed that it needed to protect the interests of Canadian manufacturers and consumers from the volatility of the world market. The sharp rise in oil prices did not reflect new production costs. As the government struggled to balance its figures, the private sector reaped the benefits of a bull market. The annual rate of return for oil and gas producers on the Toronto Stock Exchange (TSE) between 1974 and 1979 was 36.3 per cent compared to 7.6 per cent for all other firms on the TSE 300 Composite Index.[38]

Clearly, from Ottawa's point of view, the oil industry could afford to absorb costs much more readily than consumers. Provincially, the Lougheed government simply wanted to raise prices for a commodity that Albertans could sell on the world market at much higher prices. But with a new majority government in place and the public viewing Alberta as greedy, Ottawa did not have to negotiate. The industry was not held in particularly high esteem by Canadians east of Alberta. Indeed, in the early fall, a poll commissioned by the CPA revealed that 21 per cent of Canadians believed that energy was the most important issue facing the government; only inflation (24 per cent) ranked higher. More than half of Canadians surveyed thought the country would suffer energy shortages the following year and more than half indicated that they were willing to pay more if it meant guaranteeing energy supplies. More alarmingly for the CPA was that over 75 per cent of Canadians favoured increasing Canadian control and ownership of the oil industry through government intervention.[39]

Whether by design or inertia, the oil industry and its representative organizations, like the CPA and IPAC, lost the power to define the terms of debate. Trudeau casually denied repeated attempts by Lougheed to jump-start the talks again, including a public pledge by the provincial leader to fly to Ottawa on a moment's notice.[40] In the waning days before the budget came down, Trudeau did not need to negotiate, he could afford simply to bide his time. The soon-to-be-announced National Energy Program would accomplish Trudeau's goals without the need for negotiation and would give Ottawa control of the political agenda.

Significantly, the study that initiated the National Energy Program began as a paper in the Department of Energy, Mines, and Resources (EMR) during the brief Clark government in 1979–80.[41] As Prime Minister, Joe Clark resurrected EnFin, a group of high-level bureaucrats from the departments of Energy and Finance, to co-ordinate the federal government's position for negotiations with Peter Lougheed.[42] The sheer number of potential political and commercial interests, combined with the recognition that competing departments were prepared to resist any energy plan, made secrecy imperative. Complete secrecy, however, was impossible due to the extensive policy review process unofficially called the Pitfield system.[43]

The Pitfield system was a carefully constructed and conceived "technocracy" created by Trudeau's Liberals as a means to set long-range priorities, requiring interdepartmental consultation and imposing cost-benefit evaluation. Trudeau and his ministers originally used the system to consolidate the Liberals' control over the bureaucracy from 1972 to 1979. In 1980, they also realized that they could use the system to further their agenda. Often, "ambitious and imaginative legislative proposals could suffer dissipation of momentum and dilution of substance" as they made their way through tedious departmental reviews. The new use of the Pitfield system would keep the bulk of the government's energy proposals intact.

Partly as a response to the need for secrecy, Clark exempted EnFin from submitting policy drafts and budget proposals through the Cabinet altogether. Instead, the group would report directly to the Prime Minister's inner cabinet, limiting the number of different observers and institutional checks on policy. Trudeau liked the general structure of EnFin and kept it intact for use by Lalonde, augmenting the minister's power by awarding EnFin its own multibillion-dollar budget "envelope."[44] Therefore, despite going through eighteen different drafts of legislation, the details of the NEP remained a closely guarded secret known to few ministers and officials within the government itself.

A week before the October 28 budget was read, the TSE Oil and Gas Index peaked at 5073.40 points before pre-budget jitters trimmed it back to

4882.31.[45] Many drilling and oilfield service companies announced tentative plans to cut expenditures. Exploration departments braced themselves for mild cutbacks. Production estimates for 1981 anticipated a decline of approximately 8 per cent – roughly 100,000 barrels per day – at Alberta refineries. Provincial officials nervously commented that many companies were poised to move their operations to the United States if the federal government moved too vigorously against the industry.[46]

The Calgary Herald captured the industry's gloomy mood on the day the budget came down. "The first real shots are about to be fired in what could prove a bitter, protracted legal and economic war between Ottawa and the province of Alberta," predicted one writer. "The action begins tonight when Finance Minister Alan MacEachen delivers what is expected, at least in Alberta government circles, to be a hard-hitting budget."[47] Another analyst speculated that a move against provincial oil and gas revenues by Ottawa would provoke a constitutional crisis. The resulting political instability would adversely affect investor confidence in the Alberta oil industry. Years later, the former president of IPAC, Gwyn Morgan, recalled the uncertainty swirling in business circles. "We knew we were going to get hit," remembered Morgan. "But we had no idea that we'd be singled out, that this was not a budget at all, but an energy policy that would change the whole oil industry."[48]

At eight o'clock on October 28, 1980, the Canadian House of Commons reconvened to hear the Liberals' first budget since returning to power eight months before. Allan MacEachen rose and began to present the government's budget ten minutes later. "It would be no service to this House," began MacEachen, "or to Canadians, to deny that there is a deeply troubling air of uncertainty and anxiety around the world and, I am sure, in the hearts and minds of Canadians." MacEachen painted a stark picture of the world situation before turning to the substance of the budget. Not surprisingly, the Finance Minister turned first to the government's new energy policy. "Time is running out," said MacEachen emphatically. "The federal government feels compelled to put Canada's energy house in order."[49]

Nearly an hour later, the Finance Minister finished outlining the basic parameters of the National Energy Program and quickly presented the rest of his budget. A thick booklet, appropriately titled *The National Energy*

Program, which accompanied MacEachen's budget, provided a more detailed description of the government's policy. Some analysts suggested that the authors of the NEP designed a political program to reign in the oil industry, rather than producing a sound economic policy.[50] The document's polemical introduction, personally authored by Marc Lalonde, fired the first shots across the bow of Canada's oil industry. "In Canada, one provincial government – not all, and not the national government – enjoys most of the windfalls under current policies," argued Lalonde. "These policies are no longer compatible with the national interest." American multinationals dominated the Canadian oil industry, and the government asserted that American interests owned between 70 and 80 per cent of Canadian petroleum firms.[51] Thus, Lalonde argued that the NEP would try to achieve three goals:

1. To establish the basis for Canadians to seize control of their own energy future through *security* of supply and ultimate independence from the world market,
2. To offer to Canadians, all Canadians, the real *opportunity* to participate in the energy industry in general and the petroleum industry in particular, and
3. To share in the benefits of industry expansion; to establish a petroleum pricing and revenue-sharing regime that recognizes the requirement of *fairness* to all Canadians no matter where they live.[52]

To achieve these goals, the NEP proposed massive government intervention into the nation's economy. This was the first attempt since the Second World War to mobilize the nation's resources. Lalonde offered detailed regulations and plans to enable Canada "to dissociate itself from the world oil market of the 1980s."[53]

The first target was crude oil pricing. Since the early 1970s, the federal government argued consistently that the world crude price bore little resemblance to actual production costs. OPEC's decision to control production and drive up world prices reflected political, not economic, considerations. Therefore, since Canadian crude cost less than world prices, the federal government ensured that Canadian producers recovered their costs plus some profit.

Producers could not, however, expect the "windfall profits" other international suppliers accrued by charging the world price. Higher world prices made feasible the exploration of more "marginal" areas, like the Arctic and the Beaufort Sea; Ottawa wanted to encourage such exploration as well. Lalonde declared that oil discovered before 1979 was "old" oil, and therefore should be cheaper in price than "new" oil – that is, oil discovered after 1979. Rather than charging two prices – one for "old" oil and one for "new" oil – Ottawa would charge Canadian consumers a single "blended" price. By "blending" the price of "new" oil with "old" oil, the net result would be cheaper prices for consumers.[54]

As for synthetic crude projects, the NEP acknowledged that development was not profitable unless they charged a higher price, but it insisted this "need not be as high as the international price." The NEP dictated that the practice of selling synthetic production at world prices would stop immediately. Instead, the NEP contained provisions to establish the New Oil Reference Price (NORP) for synthetic projects that would result in a "made-in-Canada" price for oil sands and heavy oil projects. Although officially tied to the world price, the NORP still established prices below world levels and applied specifically to heavy oil projects and oil discovered after 1974.

Perceptions prevailed in Fort McMurray that the NORP section specifically targeted Suncor. By virtue of an agreement between the company and the federal government, Suncor had charged world prices for its full production since April 1979. Higher prices enabled the company to offset costs incurred while expanding the facility. "The revenues accruing under this agreement have more than covered the expected capital costs of the expansion, and unwarranted windfall gains would result if the arrangement were continued," warned the NEP. From now on, Suncor would only charge the lower NORP.[55]

Although Lalonde's methods seemed arbitrary and capricious, nothing was coincidental about them. The Canadian oil industry and the government of Alberta immediately understood Ottawa's strategy. According to the National Energy Board's 1980 Annual Report, the oil industry had already developed 65 per cent of Canada's known crude oil reserves. Alberta still possessed 83 per cent of the country's remaining 35 per cent.[56] To achieve energy security, Lalonde wanted oil companies to develop marginal oilfields. Not only would this secure Canada's oil supply, it would entice producers onto Crown Lands (also known as the "Canada Lands") – in the Arctic, Atlantic and Pacific Oceans – and away from the Province of Alberta[57] (Map 2).

The federal government proposed to use tax policy to encourage the shift from provincial jurisdiction to federal lands. The NEP proposed to eliminate the industry's earned depletion allowance for wells drilled outside the Canada Lands. To encourage Canadian development of Canadian reserves, the federal government introduced Petroleum Incentive Payments (PIP). These favoured only *Canadian* producers who developed wells on Crown Lands. To qualify for the government's new incentive program, oil companies had to be able to prove that Canadians owned more than 50 per cent of the company. In any case, the NEP would bar large foreign-owned multinationals from developing oilfields on Crown Land.[58] Lalonde's chosen champion for "Canadianization" was Petro-Canada, and the National Energy Program included a controversial "back-in" provision to ease Petro-Canada's expanded role. Through the "back-in" clause, Petro-Canada would automatically obtain a 25 per cent ownership in every new oil development project that occurred on Crown Lands. The federal government announced that it would appropriate capital to Petro-Canada to buy out more foreign-owned corporations.[59]

By themselves, these provisions would likely provoke a visceral response from the oil patch, particularly Petro-Canada's status as the chosen arm of the federal government. Nevertheless, Lalonde also introduced a range of new taxes to divert billions of petrodollars away from provincial treasuries into federal government coffers. The Petroleum and Gas Revenue Tax (PGRT) instituted a wellhead tax on petroleum products as they came out of the ground. The government also taxed gasoline as it left the refinery and earmarked these funds to pay for Petro-Canada's foreign acquisitions with the "Canadian Ownership Charge." At the pump, consumers would pay a tax on gasoline to help finance the federal government's subsidy on imported oil.

Clearly, the NEP inaugurated a new era in federal-provincial relations and made a comprehensive energy policy that aimed a dagger right at the

Map 2   Canada Lands

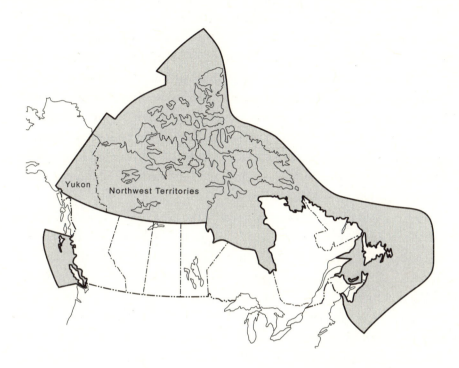

heart of the oil industry. John Crosbie, the Opposition's chief financial critic, wasted little time and blasted MacEachen's budget on the floor of the House of Commons. "The minister has a majority in this House," said Crosbie. "He has not got the courage of a castrated lemming." [60] Privately, Crosbie greeted the news of the NEP with barely concealed smugness. Peter Lougheed had finally received his comeuppance for failing to negotiate in good faith with Clark's Conservative government. In his memoirs, written a decade later, Crosbie did not recall the catastrophe that hit the oil industry, nor did he focus on the creation of the NEP. Instead, Crosbie painted the picture in the starkest of political and personal terms. A year before Lougheed had helped force the federal Tories from power. Now Trudeau was cutting the rest of Alberta's "blue-eyed sheiks" down to size.

The National Energy Program exploded on Alberta like a bomb. The editorial writers for *The Calgary Herald* opined that the NEP was "incredibly lofty and patronizing for a government document in a democratic country."[61] "This," concluded energy analyst Charles Lynch, "is what happens when you get a wily old recluse as your finance minister." [62] More damning than the words of editorial writers, however, was the reaction of Canadian investors. Within two days of its announcement, the TSE oil and gas index dropped more than 800 points. Roughly $2.3 billion of investment money and approximately 16 per cent of the index's total value disappeared as investor confidence disintegrated.[63] Stock brokers dumped shares in Canadian oil companies so quickly that the TSE suspended trading on several petroleum stocks. C.J. Lawrence Incorporated, an American brokerage firm specializing in oil and gas stocks, issued a blanket "sell" recommendation on shares for all Canadian petroleum companies.[64]

Grim-faced C.E.O.'s were stunned by the NEP. "This seriously appears to be a uniformly dismal budget," said Jack MacLeod, president of the Canadian Petroleum Association and Shell Canada Resources. "The more we go into details the worse it gets." [65] "It's incredible," said Gordon Ritchie of Dominion Securities. "There is absolutely no economic incentive to go out and explore for new oil and gas." Alan Jones of Pemberton Securities concurred with Ritchie and added that the budget would simply encourage the flight of capital to the United States.[66] Bumper stickers bearing the slogan, "Let the Eastern Bastards Freeze in the Dark!" reappeared on a few Alberta vehicles, reflecting popular hostility towards Eastern Canada.

Compared with the rest of the industry, the oil sands apparently escaped from the NEP unscathed. The oil sands industry managed to retain its depletion allowance, obtained higher world prices for its output, and generally received special consideration from the federal government. "This program offers high, rising and predictable prices," argued the NEP's authors. "The reference price for the oil sands is *essentially equivalent* to the current world price, and will be escalated by the inflation rate." Provided the Province of Alberta approved the Cold Lake and Alsands projects and established a "reasonable" royalty structure, the federal government saw "no further reason for delay in construction of these important projects." [67]

Suncor officials begged to differ with Ottawa's interpretation. Although world prices held around $38 per barrel, Ottawa established the reference price for oil sands at $16.75 per barrel, approximately 44 per cent of the international price. The details of the new pricing scheme made a complete mockery of the federal government's claim that the oil sands reference price and the world price were "essentially equivalent." The NEP, concluded Suncor vice president Dudley McGeer, was a slap in the face to the synthetic industry. "Not receiving world price will knock $1.5 billion out of the projected cash flow over the next five years." McGeer pointed out that the new price would not even cover Suncor's 1980 operating costs. By rolling back prices and slapping the company with an industry-wide 8 per cent petroleum tax, the economics of oil sands operations received a staggering blow.[68]

For two days, Premier Peter Lougheed remained silent about the federal budget and the NEP, declining repeated opportunities to comment officially. Other government officials and commentators speculated that the premier would act soon. Some believed that Alberta would "turn off the taps" and deny Eastern Canada any of Alberta's petroleum, and would hold up approval for oil sands projects indefinitely.[69] In a tense and combustible political atmosphere, Lougheed completed the province's response on October 30, and, amid tight security, recorded a prepared statement to Albertans broadcast later that night. In a firm voice, Lougheed asserted that Albertans were "entitled to respond in a measured way" to the National Energy Program. "I think what's happened is that the Ottawa government has, without negotiation, without agreement, simply walked into our home and occupied the living room."[70]

Instead of accepting the newly imposed *status quo*, Lougheed announced that Alberta would cut sales of petroleum products to the rest of Canada by 15 per cent, or 180,000 barrels a day. Alberta would carry out the measure in three equal cuts spread out over nine months. Lougheed also announced that the Alberta government would put a hold on the development of the Alsands and Cold Lake projects. The premier surmised that these projects would serve as "trump cards" in negotiations with Ottawa because of the importance of the oil sands in securing Canadian self-sufficiency by the 1990s.[71] Despite Lalonde's discussions about Hibernia and developments in the Beaufort Sea, Lougheed believed Ottawa needed the oil sands to replace declining conventional supplies. Finally, the province announced that it would "challenge [the natural gas export tax] legally because we believe we have a reasonable case." [72]

Lougheed intended the October 30 address to serve as a starting point for discussions between the federal and provincial governments. Fortunately for all sides, this is exactly how Trudeau received it. In the Prime Minister's eyes, Lougheed was simply a tough negotiator who was looking out for the best interests of his province. [73] Energy Minister Marc Lalonde, however, viewed Lougheed's speech in a much more hostile light. Lalonde claimed the move by the Alberta Legislature to cut petroleum production and to hold up approval of the two oil sands plants amounted to a clear threat to "the national interest." In an interview with the Southam News chain, Lalonde suggested that the federal government might eventually resort to emergency authority to seize control of the country's oil and gas supplies. [74]

REACTION OF THE OIL INDUSTRY

As the battle between Edmonton and Ottawa over the future of oil and gas revenues heated up, few bothered noticing that discussions between Lougheed and Trudeau were moot. The oil industry had already initiated a contraction that would made subsequent discussions pointless.

The increasingly hostile investment climate in Canada, brought about first by the NEP and then by Lougheed's response, caused many oil companies to divest themselves of their Canadian holdings. After the federal government revealed the details of the NEP, many multinational oil companies moved their equipment to the United States before December 1980. The December 15 issue of *Oilweek* noted that companies had already "stacked" two hundred rigs (non-productive) or were prepared to leave the country altogether. [75] Deregulation of the U.S. oil industry gave the Americans a more attractive market compared with Canada, and most companies simply hauled their equipment and investment dollars down south. [76] Before the NEP announcement, the Royal Bank of Canada estimated that foreign sources needed to provide $350 billion in investment capital to finance all Canadian energy programs. [77] During the latter years of the 1970s an average of $2 billion left the country annually. But in the year after Lalonde unveiled the program, capital flight rocked the Canadian economy by roughly $1.2 billion Canadian *per month*. [78]

For oil sands producers, the situation deteriorated. Double-digit inflation and interest rates, skyrocketing material and labour costs all raised start-up costs exponentially. The partners in the Alsands consortium began to look uneasily at the fiscal bottom line. A re-evaluation of the project in April 1981 showed that start-up costs more than doubled, from $5.1 billion to $13 billion.[79] Given the current tax and royalty regime, some consortium members began to re-evaluate their commitment to the project. Under the NEP, Alsands would pay 8 per cent of the PGRT until project completion and then 12 per cent after that. These totals were beyond provincial income taxes between eleven and 15 per cent and the 30 per cent royalty levied by the provincial treasury. Because the consortium had several different multinational corporations, the new tax structure established by the NEP affected each differently.[80]

The Alsands partnership group began expressing concern over the yet unspecified federal pricing scheme. Taken together, these factors began to shake the partners' interest and faith in the project. It was one thing to invest in a synthetic oil facility with all its attendant risks. It was another to invest $13 billion in a project subject to the whims of two governments locked in political combat.

In early 1981, political and public forces began pushing Edmonton and Ottawa toward a settlement. Lougheed's production cuts did not have their desired effect. Oil was available on the world market to make up the loss of Alberta's production. Perhaps more importantly, Alberta was losing the public relations battle because of the production cuts. The federal government capitalized on this move by instituting what it called the "Lougheed Levy" on consumer prices to pay for additional imports. Business leaders, including the Canadian Chamber of Commerce, expressed their frustration at the lingering impasse. Even the CPA's patience was beginning to thin. CPA president J.M. Macleod suggested that Lougheed had become too greedy in his quest for additional provincial royalties. "Both governments must bend to provide greater revenues to the companies." [81]

As the province and the federal government began negotiating, a radical change had clearly taken place. In the 1970s, the province sat in the driver's seat and effectively determined what Canada's energy policy would be. Now the federal government seized the initiative and forced both the province and the industry into the unaccustomed role of responding to federal proposals. But the industry found itself excluded from discussions as the province prepared its response. Finally, the province also began negotiating with the federal government without prior consultations with the industry. Although uneasy at the turn of events, the mood in Calgary's corporate offices remained optimistic: Alberta's interests were the industry's interests.

When the province and the federal government emerged on September 3, 1981 with a new pricing agreement, many in the industry automatically believed Peter Lougheed had negotiated a good settlement. Bill Daniel, the chief executive of Shell Canada, even praised the deal before examining the details. However, after he had a chance to read and digest the document, the results were far from encouraging. "Company after company had put the agreement's numbers into their computers and came up with similar results," wrote Foster. "Far from returning the industry to a state of pre-NEP health, the deal didn't pan out to be much better than the NEP itself." [82]

The bitter lesson the industry learned is that corporate interests were different from provincial interests. The Agreement retained the basic provisions the NEP outlined for the oil sands. The NORP emerging from the federal/provincial agreements established a higher price for synthetic crude and pegged it at 85 per cent of world prices. By virtue of the pricing agreement, the NORP rose to $38 per barrel and the Consumer Price Index determined subsequent increases. The only problem was that this new price would not be enough to ensure that synthetic producers would realize a significant rate of return on their investment. A subsequent Economic Council of Canada paper argued that the APG's costs for producing a barrel of synthetic crude were between $37 and $48 – before they paid royalties and taxes to both levels of government. In all likelihood, the Economic Council of Canada argued, the partnership group would lose at least $2 per barrel. [83]

Beyond this consideration, the price level established by the NORP for synthetic fuel projects would only remain in effect until 1986. [84] At that time, both levels of government pledged to sit down together to re-examine and renegotiate the NORP for synthetic projects. For any oil sands project looking to attract investment dollars, the uncertainty created by NORP served as a disincentive to new project start-up. As entrepreneurs, oilmen could accept that fluctuations in market prices could make or break their investments. But they did not trust that a politically inspired price control would grant a reasonable rate of return.

Other problems remained for oil sands producers, particularly regarding the tax structure. The new Agreement stipulated that the base rate of the PGRT doubled from 8 per cent to 16 per cent for conventional producers. Meanwhile, oil sands producers learned that 75 per cent of their revenues were now subject to the Incremental Oil Revenue Tax (IORT) of 50 per cent.[85] Together, these two provisions effectively placed a ceiling on the industry by limiting available capital. Oil sands producers argued that the new taxes established a "ring fence" around the synthetic industry. The "ring fence" limited the revenue that companies could shield from income taxes, and hurt development. In this regard, the new taxation provisions did not take into account the interrelationship between the conventional and synthetic industries. Since synthetic projects required large amounts of start-up capital, companies relied on profits generated from the conventional side of the industry to fund synthetic projects. The new tax regime could not cleanly separate the two.[86]

Nothing suggested to the oil industry that Trudeau's new policy was simply a passing phase. In fact, with each passing day, the determination of the federal government to see its program through increased even as the prospect of achieving energy self-sufficiency grew more remote. With each round of layoffs and cutbacks in exploration budgets, Canada's domestic production decreased. In retrospect, the extent to which the Canadian government believed it was possible to divorce the Canadian oil industry from international markets remains the most troubling aspect of the NEP. If government projections about the world price had been accurate, the NEP would have resulted in substantial benefits for Canadian consumers. However, by divorcing itself from the world market so completely and so thoroughly, Canada also divorced itself from reality. Ottawa removed the framework capable of establishing the true value of synthetic crude projects and substituted its own.

Moreover, the "Canadianization" provisions of the NEP discriminated against American multinational corporations. Several American firms, beginning with J. Howard Pew's Sun Oil, had made significant contributions to the research and development of oil sands operations. Now, federal government policy drove them and their investment capital back across the border. Prospects for oil sands development in the 1980s, therefore, took a cruel turn few had foreseen. Just as government was inclined to embrace oil sands development, it set up a pricing, taxation, and royalty regime that ensured no business could afford to develop them.

The cruellest twist of all took place just after Ottawa decided to turn its back on the world petroleum market: the world petroleum market made itself impossible to ignore. High OPEC prices – between $36 and $42 per barrel attained in the aftermath of Iraq's attack on Iran on September 23, 1980 – were unsustainable. The high prices began a recession throughout the industrialized world that, in turn, decreased the overall demand for crude oil and oil products. Multinational oil companies also learned their lesson well after the outbreak of the Iranian Revolution; they spent the intervening years storing supplies. "The Great Panic [of 1979] had, by its own logic, turned into the Great Inventory Build," wrote Yergin. "And when the war broke out, storage tanks all around the world were brimming over." [87]

Increased supplies from Saudi Arabia, Mexico, Britain and Norway more than offset the loss of Iranian and Iraqi production. Non-OPEC producers, anxious to capitalize on the situation, began to sell their production below world prices, reducing demand for OPEC oil by 27 per cent. [88] Spot oil prices, instrumental in forcing world prices higher four years earlier, fell below official prices. "Even a war between two of the most important exporters could only retard but not cancel out the powerful forces set in motion by the two oil shocks," concluded Yergin. "October 1981 represented the last time that the OPEC price would go up, at least for a decade." [89]

The drop in world prices bore out Ottawa's contention that world crude prices bore little resemblance to the actual cost of production of conventional sources. However, Ottawa's new problem was that, given the state of technology in 1980–81, the artificially high world price *did* reflect the costs of producing synthetic fuels. The higher the world price rose, the more feasible synthetic operations became. This reality led Peter Foster to observe that synthetic producers were in the "OPEC Paradox." With dwindling conventional supplies, Canada was dependent on the expensive oil locked in the oil sands and in the frontier projects described in the NEP. Those projects, however, were entirely dependent on receiving world prices for their output to make them economically feasible.

> To the extent that these projects are developed, and Canada, and other countries, succeed in weaning themselves away from OPEC oil, then OPEC oil will be in increasing surplus and its price will drop. This in turn will make the synthetic oil and frontier projects uneconomic. To the extent that these projects are *not* developed, then OPEC oil will be in increasing shortage and its price will rise. This in turn will make the synthetic oil and frontier projects economically attractive.[90]

The combination of the drop in world prices and the reduced incentives offered by the federal and provincial governments radically scaled back the private sector's interest in synthetic fuel projects. The industry abruptly cancelled many oil sands ventures when world prices nose-dived in early 1981. Imperial Oil announced in July 1981 that it would indefinitely suspend its Cold Lake Project and Syncrude shelved plans to expand its operations. The Alsands Partnership Group collapsed despite a last-ditch attempt by the province and the federal government to save the project on April 30, 1982.[91]

The collapse of the Alsands group was, perhaps, the bitterest pill for both Edmonton and Ottawa to swallow. Neither government appeared to take seriously the notion that their restrictive policies might force the cancellation of the project. Disbelief turned to panic in early April as Edmonton and Ottawa cobbled together a new incentive package to salvage the project. Edmonton pledged a loan guarantee equal to 20 per cent of private sector investment in the project. Ottawa reduced the PGRT to 0 per cent until the partners turned a profit. Two of the most financially sound companies, Amoco and Chevron, still considered the deal too risky and dropped out. Within a matter of days, Shell Explorer and Dome Petroleum also indicated they would not participate in Alsands. Further government concessions contained in the "final offer" – government ownership of 50 per cent, increased federal and provincial loan guarantees, and a postponement of royalties and taxes – did little to alter the demise of Alsands.[92]

The Alsands partners were not looking to renegotiate the terms of the deal. Canada's new investment climate made large-scale oil sands projects too risky for private enterprise to undertake. After final attempts by the federal and provincial governments failed, all sides stood back for a second sober look at a failed development policy. "The day Alsands died was a tough day," recalled Lougheed. "In fact, one friend of mine said the last day of April 1982 was one of the worst days politically in a rather dismal time because we'd gone through the downturn of the 1981–82 winter; we'd had a separatist chosen in a by-election in February – then came the crusher, Alsands collapsed."[93] Competing and conflicting government interests had doomed projects important to Canada's future. Defining the province's oil sands policy throughout the 1960s and early part of the 1970s was relatively easy and straightforward; the same was not true by the early 1980s. Instead of presenting concrete initiatives, both levels of government remained content to mouth platitudes about the importance of the oil sands while passing legislation that inhibited their future development.

A vicious cycle trapped oil sands developers where they could afford neither to go forward nor turn back. "Given current technology and the implied capital and operating costs, the short-term price outlook, and no guarantee of stable if not escalating real prices for tar sands oil, it is unlikely that any further plants will be built," concluded three Queen's University academics in 1982.[94] Without future plants coming on-stream, the incentive to develop cost-cutting technologies evaporated and limited the possibility that the synthetic industry could evolve any further. "Since capital and operating costs of current technologies will doubtless continue to rise," concluded the Queen's study, "tar sands plants could continue to be uneconomic even if real world oil prices do rise substantially."[95]

Finally, in a speech in late August 1982, Alberta's Deputy Energy Minister, Dr. Barry Mellon, acknowledged what many in the oil industry already knew: government was responsible for many problems. In the aftermath of the Alsands debacle, government must furnish investors with the "rules of the game" that would apply to each project, including royalty and tax regimes. Mellon challenged the federal government to assume its share of the responsibility for the failed policy. The oil sands must have access to world prices "unencumbered by qualifications, caveats and footnotes which have become the hallmark of the Ottawa bureaucracy. If Ottawa is prepared to pay a premium price for foreign crude," concluded Mellon, "it must allow for an equivalent price for domestic, including synthetic." [96]

Encouraged by Mellon's statements, the CPA prepared a position paper on the future of oil sands development in Alberta late in 1982. The paper showed that the CPA's membership agreed, mostly, with the statements made by Mellon in his August speech. The CPA's reply returned to basic themes. Any new oil sands policy must state its goals clearly from the outset and apply equally to all developers. "The case-by-case approach and secret negotiations do not provide a viable investment climate and should be undertaken only in cases where modifications or alterations are in the public interest," argued the CPA. The association pointed out that the industry required a large initial investment and thus represented a considerable risk for any operator. The CPA suggested that excluding oil sands production from pro-rationing measures and allowing synthetic producers to charge world prices would reduce some risks. Meanwhile, project owners should be responsible for developing their own markets to avoid conflict with conventional producers. Finally, the federal government should reconsider its position on exports and allow producers to sell excess production abroad. [97]

What is particularly striking about the arguments put forward by the CPA is how closely they mimic arguments that oil sands producers advanced over the prior two decades. Perhaps after a decade of confrontation and frustration with both the federal and provincial government, the industry was simply getting reacquainted. On the other hand, perhaps they simply prove the truth of the old maxim that interests never change. Regardless, it took a long time for the CPA to find a receptive audience. As it turned out, the basic provisions of the NEP remained in place until March 1985 when the newly elected federal government of Brian Mulroney dismantled the program. [98]

In 1980, the National Energy Board projected that between two and five oil sands plants would be constructed by 1995. Unless the oil sands became more economically viable, the industry doubted a single project would be completed before 2000.[99] Nonetheless, by the mid-1980s the oil sands began to make a tentative comeback by scaling back the size of the industry's operations; gone were the "mega-projects" envisioned in the late 1970s. In their place, companies designed flexible projects they could tailor to market conditions. Imperial Oil, for example, scaled back its Cold Lake project by proposing a modular facility and dropping plans to include a heavy oil upgrader in its operations. The ERCB approved the plan in 1983 and Imperial began production at the plant in 1985. The provincial government made timely loans and modified the royalty regime to ensure that Syncrude completed its proposed $1.5 billion upgrade of plant facilities in 1988.[100]

Once again, however, the world market intervened to bring progress on the oil sands to a screeching halt. The world price of crude oil would drop sharply in 1986 when OPEC's member nations continued to cheat on their production quotas. Non-OPEC producers exacerbated the problem by continuing to pump more oil out of the ground than they could sell. Oversupply, not scarcity, defined the market for the remainder of the decade, making the economics of the oil sands even more untenable than they had been during the NEP. Prices dropped from a high of $31.75 a barrel in November 1985 to $10 a few months later.[101]

In effect, the NEP gutted the Alberta oil patch and severely jeopardized the country's energy future. As with other industries, the pace of the market determined the nature and course of the petroleum industry's development, and one leading indicator of industry health is exploration. In periods of expansion – when prices are high and revenues increase – companies direct profits toward exploration. When the industry is at the top of a business cycle, exploration budgets are typically high to increase expenditures and to take advantage of tax write-offs. However, when the industry enters a recession, the industry contracts and eliminates speculative spending. Since the emphasis is on consolidation and maintenance, exploration budgets are normally among the first sectors cut back. The NEP, therefore, resulted in a counter-cyclical swing, and deprived the oil patch of revenues necessary to sustain continued growth and expansion, particularly in the oil sands.

The Canadian Petroleum Association later blamed the NEP for the loss of 15,000 jobs, a 22 per cent drop in drilling activity and a 25 per cent decline in exploration budgets. Overall, industry cash flow and earnings fell

34 per cent to $3.1 billion in 1981 from $4.7 billion a year before. The price paid by consumers of petroleum products almost doubled as consumption dropped by 6 per cent.[102] The industry realized far too late that the energy struggles between the province and the federal government had little to do with oil and everything to do with political power. As Bruce Doern and Glen Toner convincingly argue, Lalonde designed the NEP from the very outset "to alter the structure of power between Ottawa and the industry as a whole, between Ottawa and foreign-owned energy interests in particular, and between Ottawa and Alberta."[103] While it accomplished those goals, it also decimated the conventional oil and gas industries along the way.

Economists would later write that the NEP cost the provincial economy more than $97 billion, but the synthetic industry paid a greater price.[104] Combined with the drop in world oil prices in 1986, the result delayed development of Canada's oil sands by five years.[105] Perhaps, as Daniel Yergin suggested, the dilemma with synthetic fuel projects is that they appealed to North American desires for a quick technological fix, but exerted very little influence over world crude supplies. At one energy conference in California in the late 1970s, oil executives regaled the audience with tales of tapping the North Slope in Alaska. Next came the nuclear power people talking about advanced nuclear technology. Synthetic fuels representatives pointed to enhanced recovery efforts, slurry pipelines, oil shale and the oil sands as potential solutions to the chronic shortage of available crude. At the end of all these dynamic presentations, an engineer from the City of Los Angeles Department of Water and Power stumped them all. He explained that closing the curtains at night could radically reduce energy consumed in the United States. The irony was palpable. "Here was a conference devoted to the great energy crisis, the moral equivalent of war," observed Yergin, "and here was a man who was saying that the solution is for people to do such bold things as close their curtains at night."[106]

Fortunately, for Canada's oil sands producers, Carter's massive Energy Security Corporation was cancelled by President Ronald Reagan.[107] Overall, the decade that began with so much promise for oil sands producers ended with few accomplishments. Yet again, developments delayed the era when Canadian crude supplies made a significant contribution to national energy supplies.

CHAPTER 8

Competition's "Cold Shower": Remaking the Oil Sands Industry in the Era of Globalization and Free Trade, 1984–2001

The principal opportunity for domestic and export sales of oil sands crude is to North American refineries, where there has been an increased capability of running larger volumes of heavy and sour crudes. Future market development might take the form of petroleum products or petrochemicals, resulting in additional facilities similar to those found in Fort Saskatchewan, provided these products can economically compete. – *The National Oil Sands Task Force*[1]

The 1990s marked a new beginning for oil sands producers. No longer intimidated by the technical, environmental, and financial challenges of the Alberta oil sands, the industry pressed to streamline procedures and to improve business opportunities for synthetics. The new agenda for oil sands evolved in response to the obstacles the industry encountered in the late 1970s and early 1980s. Government and industry gave sustained attention to three areas in the early 1990s – the regulatory environment, technological developments, and the search for markets.

Ottawa and Edmonton came to realize that the regulatory and tax environment created at the height of the "energy wars" inhibited oil sands development. The competition between federal and provincial governments

197

to gain a larger share of industry revenues had crippled development and placed the country's petroleum industry at risk. Without synthetic fuel development, Canada would rely on greater quantities of imported oil because of rapidly declining conventional crude stocks. The lesson was simple: without compromise between public and private sectors, the oil sands industry stagnated.

In retrospect, the collapse of the "mega-project," the sharp drops in world crude prices (from $34 in 1980 to $10 in 1986), and the harmful feud between Edmonton and Ottawa over natural resource jurisdiction proved a blessing in disguise for the industry. In the 1970s and early 1980s, the industry embraced "bigness" as the path to success at the expense of streamlining operations and experimentation. The trend toward larger plants, financed by government loans and sustained by tax and royalty concessions, received mortal blows from the twin shocks of the NEP and the drop in world crude prices that rendered large-scale synthetic projects uneconomical. But the final death blow for oil sand mega-projects came in 1992, two years after the federal government abruptly withdrew funding for the struggling Other Six Lease Operators (OSLO) Project.

Confronted by the possibility that the bloated nature of the oil sands industry could spell the end of Canada's synthetic fuels industry, oil sand producers searched for answers. Surmising that international commodity prices alone might not sustain synthetic development, the industry adopted new strategies and incorporated new technologies to cut costs and compete with conventional crude. Government reliance on the market hastened changes to the industry's operations. Although the term "globalization" never made its way into Alberta public debates, the process shaped the province's future. The Alberta government under Ralph Klein restructured the provincial government and signalled a philosophical change regarding the role of the government in the oil sands industry. No longer would the Province of Alberta be involved directly in oil sands projects. Instead, Edmonton would strive to create a favourable investment climate for the private sector to continue research and development. Working together, government and industry would create the conditions necessary for small-scale private sector development to continue where big business left off.

Under Premier Peter Lougheed and his designated successor, Don Getty,[2] the government continued to grow and it became less responsive to the needs of the electorate. Political scientist Brooke Jeffrey concluded that Getty "turned out to be the wrong man for the job."[3] From 1971 to 1985, provincial spending increased from $1.1 billion a year to $10.8 billion a year. The size of the civil service tripled and "Alberta's Tories got in the habit of spending money faster than it came in," wrote journalist Andrew Nikiforuk.[4] Getty's administration was rocked by ballooning deficits, increased spending and the failure of several government-sponsored ventures in the late 1980s and early 1990s.[5] Although a genial, affable man, Don Getty did not have the necessary personality to make tough decisions nor did he seem to be the leader for whom Albertans were looking. "He held no strong ideas of his own," concluded one reporter, "and appeared at times to be swayed by the last person out of the room."[6] His laid-back attitude and penchant for golf gave voters the impression that he did not understand or care about their problems.

CANADA-U.S. FREE TRADE

The era of "big government" spending on energy mega-projects did not abruptly end after Pierre Trudeau left 24 Sussex Drive, although it was mortally wounded. The formation of the new federal government by Brian Mulroney's[7] Conservative Party in 1984 marked a watershed in Canadian politics. Concern for the national economy became acute as Canadian manufacturers increasingly engaged in international competition. Traditional export markets threatened protectionist measures, and others were closed because trade became dominated by large trade blocs. Furthermore, an entire generation of politicians – including Trudeau and Lougheed – retired from politics. The time was ripe for change.

As political economist Stephen Clarkson writes, Brian Mulroney did not have a clear, coherent vision to refashion the country.[8] Mulroney's politics, argued Peter Newman, were utilitarian and opportunistic. As the Prime Minister himself explained in a year-end interview, "I am not ideologically opposed to anything unless it doesn't work."[9]

By 1984, Trudeau's "big government" solutions clearly did not work. Government bailouts of private sector companies kept marginal companies operating at high cost to Canadian taxpayers. Budget deficits increased – 1972–73 was the last year the federal government ran a surplus – to $32 billion in 1983–84. Moreover, the vaunted "Third Option" policy of the Liberals failed to bear fruit. Instead, Canada could not compete against the world's strongest economies and the United States remained the country's largest trading partner. In 1982, 70 per cent of all Canadian imports came from the United States and 69 per cent of all Canadian exports travelled south of the border.[10] Free trade with the United States seemed like the only alternative. "Alone and afraid in a world dominated by powerful trading blocs," concluded historians Granatstein and Hillmer, "Canada seemed to have nowhere to turn."[11]

Mulroney's *rapprochement* with Washington began soon after taking office. "Our message is clear," announced the new Prime Minister in New York in December 1984. "Canada is open for business again."[12] The chill in Canadian-American relations from the Trudeau era soon evaporated as the new Prime Minister made improved relations a priority. Mulroney won Washington's early approval by altering the most offensive portions of the Foreign Investment Review Agency and transforming the agency into Investment Canada. The new organization's mandate was radically different from that of its predecessor. Instead of barring foreign capital from the Canadian economy, Investment Canada recruited overseas investors. Mulroney's government also fulfilled a campaign promise to Western Canada by abolishing the hated NEP. More important, in May 1986 Mulroney and the Canadian government began free trade negotiations with the Reagan administration.

During his run for leader of the Progressive Conservative Party in 1983, Mulroney explicitly distanced himself from free trade. In one interview, Mulroney stated that "Canadians rejected free trade with the United States in 1911. They would do so again." Once in office, however, Mulroney's position softened. The outgoing Liberal government appointed Donald Macdonald to head a Royal Commission on Economic Union and Development Prospects. The Commission's report, published in September 1985, strongly urged the Canadian government to seek out closer economic ties with the United States. "There is no reason to suppose that our present confidence will be undermined by an arrangement designed only to secure a continuing exchange of goods and service with the United States," concluded the report. Although Macdonald allowed that free trade required a "leap of faith," he suggested he was prepared to go forward. "We believed that after a century of successful development of our country, the will to remain Canadian was not at risk," recalled Macdonald. "The country is not dependent for survival on the baling wire and binder twine of the customs tariff."[13]

Negotiations on the Canada-U.S. Free Trade Agreement (FTA) lasted until October 1987, when the two countries reached a final agreement. The Canadian Senate, still dominated by Trudeau appointees, indicated that they would not ratify the final agreement, forcing Mulroney to dissolve parliament and call a national election. (The Canadian Senate is composed of 105 members from across the country. The governor general appoints senators on the advice of the prime minister. Senators then serve until the age of seventy-five. Because of the Liberals' long reign of power, 1968–84, they held a large majority of senate positions and opposed the FTA.) Therefore, the 1988 election became a one-issue campaign dominated by the FTA and amounted to a national referendum about the treaty. Mulroney's Conservatives supported ratification, while the Liberals, led by former Prime Minister and Trudeau Cabinet Minister, John Turner, argued against accepting the deal.

"It was splendid national discussion of a vital issue," concluded Michael Bliss, "and compared brilliantly with the moronic presidential campaign the Americans had just held."[14] On the one side, self-appointed patriots argued that free trade would mean the end of Canada's sovereignty. On the other, the agreement received the unconditional support of the business community that feared that if the deal did not go through, the Canadian economy would crumble. In the end, Mulroney won a second straight majority, taking 170 of the 295 seats in parliament and 43 per cent of the popular vote.[15]

Canada joined in the agreement to gain guaranteed access to the United States market. The United States, on the other hand, wanted unhindered access for investment in Canadian industries, particularly in the energy sector.[16] The FTA stipulated that Canada could not establish a dual price mechanism similar to the one implemented by Trudeau in the 1970s. Perhaps the most controversial portion of the FTA was Article 904(a) – the so-called "proportional access clause." The clause ensured that Canadian suppliers had to maintain export levels to the United States proportionate to "Canadian total exports shipments of that commodity in relation to the total Canadian supply."[17] Even if Canada faces shortages in its crude oil supply, the FTA declared Canada could not limit supplies to the United States in any way. "It seems," wrote political scientist Michel Duquette, "that virtually all the bilateral obstacles for energy trade have lifted with the FTA."[18]

OSLO AND THE DEATH OF THE "MEGA-PROJECT"

Despite the shift toward trade liberalization, growing pains persisted as Canadians struggled to carry out the country's economic retrenchment in the mid-1980s. Ingrained habits competed against the nebulous and evolving theories inexorably pushing the government and industry toward globalization. Heading into the 1988 federal election, the Conservatives earmarked several billions for energy projects around the country – $2.6 billion for Newfoundland's Hibernia offshore project, $400 million for a heavy oil upgrader on the Alberta-Saskatchewan border in Lloydminster, and $150 million for a natural gas pipeline to Vancouver Island. Also included in the election promises was a $1.7 billion federal commitment for the Other Six Lease Operators (OSLO) project, the last of the oil sands mega-projects.[19]

Together, the Tories' four energy projects would replace declining stocks of conventional crude oil and form the cornerstone of Canada's energy program well into the next century. But as one anonymous civil servant acknowledged, the real motive forces behind Ottawa's mega-project fever were "jobs, regional development, a future election, and pressure to rush things so announcements can be made in time to win votes."[20] For politicians of all stripes, the allure of the mega-project proved too tempting to ignore.

OSLO began in 1981 as a joint venture by six of the eight partners in the Syncrude consortium (Imperial Oil, Petro-Canada, PanCanadian Petroleum, Canadian Occidental, Gulf Canada, and the Province of Alberta) to build another oil sands plant in Northern Alberta. OSLO's original plan would, literally, use Syncrude's blueprints and technology to build and run the 77,000-barrel-per-day facility. But the collapse of world prices and an unfavourable investment climate delayed project start-up. With world crude prices hovering around $10 per barrel in the first months of 1986, the province's conventional industry struggled to survive, seemingly precluding development of another oil sands plant.

The highest levels of the provincial government recognized the precarious position of the Alberta economy. Buoyed by decades of financial success before the NEP, Albertans did not adjust well to the dramatic economic downturn in the 1980s. Alberta's unemployment peaked at 11 per cent in 1984 before dropping to just less than 10 per cent between 1985 and 1987. Other indices showed a similarly grim prognosis. Provincial GDP shrank nearly 1 per cent in 1986 and the province faced its fourth straight year with a net migration out of Alberta.[21] Faced with the conundrum, Premier Getty and his cabinet began searching for solutions to put Alberta back on its feet. Naturally, given the province's massive oil sand deposits, attention focused on a reviving interest in the oil sands.

As a capital and labour-intensive operation, oil sands development appealed to the Getty government on many levels. Investment drove the Alberta economy in the 1970s, and embarking on a large-scale project like OSLO might bring back investor dollars and confidence to the province. Oil sands projects also employed large segments of the province's workforce, from engineers to tradesmen, and would combat unemployment. Synthetic development also offered the province a way to offset its declining conventional crude stocks. Furthermore, given the energy security component, it just might attract investment from Ottawa or another provincial government, like Ontario, interested in securing future energy supplies. Only one problem existed with Getty's plan: the private sector remained solidly indifferent to launching a third oil sands plant.

To a certain degree, oil price decontrol in 1985 influenced corporate investment strategies because the effects of decontrol favoured conventional production given 1985 world crude prices. When negotiations between the province and the federal government began in early 1985, Alberta's crude fetched approximately $35 on the world oil market. At that time, however, Canadian conventional producers received approximately $29.75 per barrel. Synthetic producers, on the other hand, received approximately $41 per barrel (prices for heavy oil and oil sands production were fixed by the NORP at 85 per cent of 1981 world prices and increased 2 per cent per year thereafter.)[22]

As world crude prices began a roller coaster journey in early 1986, the outlook for the Canadian oil patch became uncertain. Beginning from an average price of $28 in December 1985, prices tumbled below $12 per barrel in early 1986. Companies immediately slashed exploration budgets by $4 billion and over 18,000 jobs were threatened.[23] By mid-March, prices rallied somewhat to $17 before tumbling back to $12 in June. In late July, Imperial Oil, the country's largest oil company, announced its earnings fell by 33 per cent to $189 million during the first half of 1986 and brokers advised their clients to sell their energy holdings.[24] A few weeks later, some oil on world markets traded for as low as $7.35 per barrel.[25]

The boom and bust cycle of energy development frustrated Getty and other members of the provincial government. Not only did conventional drilling decline, but the drop in oil prices prohibited going ahead with capital-intensive projects like oil sands facilities. Although Getty understood private sector concern for the bottom line, the politician in him chafed at the MNC's ability to pick up and leave. Exploiting the province's oil and gas provided the basis for the MNC's wealth; certainly the oil companies could repay the province when times were bad.

Getty's viewpoint was widely held among Albertans in the 1980s. Alexander McEachern, Provincial leader of the NDP, complained in April 1988 that the big multinationals "tend to buy their oil in other places" rather than helping to develop the expensive oil sands. The rest of Canadians were little better, argued McEachern. They demanded a "made in Canada" price during the oil shocks of the 1970s and then insisted on returning to the world market when crude prices dropped. Alberta, complained McEachern, would have to grant plenty of incentives and take several risks to encourage private sector involvement in the oil sands. All Canadians, concluded McEachern, should "take part in the cost of producing oil in this country."[26]

To a certain degree, members of the federal Conservative government agreed with Getty about the necessity of another round of mega-project spending to stimulate regional development. A meeting of the federal and provincial energy ministers in late January 1987 produced no formal agreement. Nevertheless, a consensus emerged that both levels of government should try to alleviate private sector risks to encourage development.[27]

The joint statement issued at the end of the conference did not go far enough to please some members of the Getty cabinet. On March 24, 1987, Provincial Energy Minister Dr. Neil Webber announced that the province should begin constructing a third oil sands facility "not at the timing of the international oil companies or the major oil companies but the timing when the governments of Canada want these projects brought into place." If government forfeited the initiative to industry, Webber predicted a long delay before "significant" investment in the oil sands occurred. "If we don't have these projects come on stream relatively soon," warned Webber, "we will be hit dramatically with respect to loss of jobs, our trade balance will diminish significantly … and certainly regional disparities will continue to grow."[28]

OSLO therefore represented a fundamental shift in the role of the state in the development of Alberta's oil sands. Government, not industry, served as the impetus for the construction of a third oil sand plant. Rick Orman later stated that the provincial government did not want to play an active role in the economic development of sands but circumstances forced the Getty government's hand. "The capital or direction is not being taken by the private sector or the capital is not available to continue economic development in this province," explained Orman.[29]

Although the Syncrude project continued with the active involvement of three governments – the federal government, and the provincial governments of Alberta and Ontario – the project originated as a private sector enterprise. Government involvement only became necessary when Atlantic Richfield pulled out of the consortium. For critics, the early and active involvement of the state in OSLO amounted to a difference in kind rather than degree from Syncrude. Fundamentally, they argued, Syncrude's origin lay in the private sector; OSLO's, on the other hand, was clearly political.

Months before the federal government publicly pledged its support for OSLO, critics focused on the thorny question of government subsidies for the private sector. When questioned about the concessions, the provincial energy minister replied that government had to "share the risks" with the private sector. Regardless, for the province, OSLO's economics boiled down to a few simple numbers besides the projected rate of return: $4 billion capital project, 77,000 barrel per day facility, 15,000 jobs created, 9,000 in the construction industry alone.[30]

Almost immediately, sceptics derided government claims regarding OSLO's viability. Pundits pointed out that Syncrude's profit margin depended entirely on the federal and provincial decision to write-off construction costs, a provision not repeated in the OSLO agreement. Generous tax write-offs and royalty payments divided federal government economists. "The OSLO project," concluded journalist Terence Corcoran, "is only 'commercially viable in itself' if one distorts the meaning of commercially viable." Notwithstanding the favourable tax and royalty regime, OSLO required a world oil price of approximately US$30 per barrel – almost double current world prices – to be profitable.[31]

Even Ottawa's own blue-ribbon Energy Options Committee failed to see OSLO's merits. Begun by federal Energy Minister Marcel Masse in April 1987, *Energy Options – A Canadian Dialogue* served as the first public forum on Canadian energy policy since the Borden Commission in 1958. Chaired by Thomas Kierans, President of McLeod Young Weir Limited, the committee was given free reign to conduct "a comprehensive review of Canadian energy issues; an examination of the present energy situation; [and] ... attempt to identify and evaluate our options for the future."[32]

The final report of the committee in August 1988 showed that a remarkable degree of consensus guided Canadian attitudes toward energy and energy development. "Virtually all energy stakeholders welcomed the current policy thrust toward a more market-oriented and efficiency-driven energy policy," concluded the report.[33] Significantly, the Energy Options committee classified OSLO as little more than a regional "make work" project of dubious benefit. "If energy security can be achieved through adaptation and choice," wrote Kierans, "then there should be no need to subsidize mega-project or any other energy supply or demand alternatives."[34]

Economists William J. Baumol and Edward N. Wolff had made similar arguments about government subsidies nearly a decade before, during the depths of the energy crisis. Subsidies of new energy sources – mostly synthetic fuels – resulted in a net energy loss for the subsidizing government because of inefficiency. Not only did alternative fuels produce fewer units of energy, but the economy also lost the energy it could expect if industry made a similar investment to develop conventional sources.[35]

The Energy Options Committee report stands as the definitive blueprint for the market-oriented direction of the Mulroney government's energy policies. However, released near the start of a bruising election campaign, the report's emphasis on supply side economics loomed as an embarrassing contradiction. Consequently, Marcel Masse refused to comment publicly on the report, creating the illusion that the Tory government disagreed with Kierans' main conclusions.[36] Getty and other Albertans proved all too willing to interpret Ottawa's silence as a repudiation of the committee's negative opinion of mega-projects and continued to push forward.

Perhaps drawing momentum from the Energy Options report, critics targeted as OSLO's Achilles heel the number of subsidies, concessions, and loan guarantees required to sustain it. Large-scale and costly development of the province's natural resources should be left to the private sector. Environmentalists joined the debate, and argued that government subsidies to the mega-project did more harm to the environment by sustaining uneconomic development.[37] Optimists maintained that OSLO's production costs would be substantially lower than Syncrude's because of improved technology. Nevertheless, nagging suspicions remained that OSLO would never be profitable without significant government concessions.[38]

Market instability in 1989 forced several conventional firms to cut exploration budgets and slash jobs. Fluctuating crude prices also focused attention on OSLO's questionable economics in corporate and government boardrooms as the partners planned their budgets. In OSLO's original letter of intent, the partners agreed that they could drop out of the project before 1991 if world oil prices or inflation made the project unfeasible. With the deadline less than twenty-four months away, two of OSLO's most important backers, Imperial Oil and Petro-Canada, questioned their commitment to OSLO without additional government subsidies to guarantee profitability.[39]

Despite the misgivings of two of OSLO's major partners and the reasoned economic arguments against it, provincial support for OSLO remained unflinching. For Edmonton, the project's importance was self-evident: production from the oil sands was necessary to offset declining conventional reserves and to ensure Canadian "energy security," a stated goal of federal government policy. However, Getty and members of his government never grasped the notion that their definition of "energy security" differed from that of Ottawa. For Prime Minister Mulroney and federal Energy Minister Jake Epp, energy security meant security of *supply*. Accordingly, federal officials believed they could best achieve secure supplies through the operation of a healthy, competitive market, not government intervention.[40] For large segments of Albertans, however, "energy security" meant nothing less than attaining *self-sufficiency*. The difference between the two positions proved more than semantic, for Alberta's definition required Canada to supply enough energy to cover its own consumption. To accomplish that task demanded large-scale development of the province's oil sands reserves.

Part of Alberta's dogged pursuit of energy self-sufficiency reflected its status as owner and repository of the country's largest natural resources. Yet fear of the international market also informed Getty's politics. He feared that low commodity prices could cripple Alberta's economy and drive the industry into the arms of Middle Eastern suppliers. High prices could prompt another wave of federal intervention similar to the NEP. Albertans feared that, ultimately, OPEC's power over world crude supplies and pricing amounted to a veto over Alberta's economic future.

In his public statements concerning OSLO, Getty repeatedly referred to an omnipotent OPEC cartel capable of destroying western economies with crippling price increases or dramatic supply shortages. OPEC, according to the Premier, was always metaphorically strangling the western world – placing a "noose around the western world," or putting its "hands around the neck of western economies," to name but two examples. OPEC's threatening presence was so clear to Getty that he could not fathom that others did not see it as well. When a report by the Conference Board of

Canada concluded that federal government would delay OSLO until 1992, Getty upped the ante. Responding to Opposition jibes, the premier pledged OSLO would not be delayed. "There's only one reason [OSLO] is there and proceeding," fumed Getty, "and that's because this government went out and fought for it."[41]

But the federal government entertained its own doubts about OSLO, not the least of which concerned its inherent contradiction with the market-driven orientation of its energy policy. Under Finance Minister Michael Wilson, the Mulroney government intended to cut expenditures to rein in the growing federal deficit. Committed to privatization, the Tories found unappealing the idea of sponsoring four large mega-projects while cutting social services. Perhaps Ottawa could sacrifice OSLO after all. But Edmonton ignored signals that the federal government would reconsider its commitment to the OSLO project in the 1990 budget. On budget day, Wilson announced Ottawa's withdrawal from OSLO. "In light of the economics of the project and the present fiscal environment," explained Wilson, "the government will not proceed with the offer of assistance for the construction of the OSLO oil sands project." Wilson tried to cushion the blow by stating that the government was "taking a more businesslike approach to assistance to business … this will place the emphasis more clearly on investing in economic development than subsidizing the private sector."[42]

On a post-budget trip through Alberta, designed to soothe ruffled feathers, Federal Energy Minister Jake Epp emphasized that the decision to withdraw from the project did not reflect negatively on OSLO's merits. Instead, the decision reflected the new federal emphasis on fiscal responsibility.[43] Loathe to acknowledge the project's economic shortcomings, few Albertans believed Ottawa's pronouncements about fiscal responsibility. Instead, most noted that federal support for the offshore Hibernia project in Newfoundland was simply a case of the East sticking it to the West again.

Wilson's budget also contained the surprise announcement that the federal government would privatize Petro-Canada, long the bane of free-market capitalists in Calgary and one of OSLO's major stakeholders. Even this piece of news came with a sour taste. At a press conference held the day after the announcement of the federal budget, Bill Hopper, Chairman of Petro-Canada, hailed the privatization decision as "a real shot in the arm for our people in Calgary, Toronto, and elsewhere." But when queried about OSLO, Petro-Canada's chairman voiced his long-standing reservations about the economics of oil sands ventures and labelled the project "a dog." "From our standpoint," explained Hopper, "OSLO did not have an acceptable rate of return. It's prudent to delay it now."[44]

The combination of Ottawa's pull out of OSLO and Hopper's ill-timed and ill-advised commentary on the province's pet mega-project left few smiling faces in Edmonton. Premier Getty and Energy Minister Orman both denounced the perceived "serious errors of judgement" contained in the federal budget, particularly involving funding decisions for OSLO. OPEC's growing strength, warned Getty, would soon have its "hands around the necks of the western world" unless funding of synthetic projects continued.[45]

Meanwhile Hopper's brutally honest assessment of OSLO's economics provoked the wrath of Rick Orman. In retaliation, Orman announced a review of all seventy of Petro-Canada's heavy oil and oil sand leases in the province. "The attitude on OSLO by the federal government leads us to the conclusion that Petro-Canada has a point of view on the project that is directly opposed to Alberta's," explained Orman. If the province determined that Petro-Canada was not "actively proceeding with development," Edmonton would yank the leases from Petro-Canada and award them to another company. "This," said an anonymous Orman aide to a reporter, "is Alberta's reaction to the federal budget."[46]

Unwilling to let the project die, the Alberta provincial government toyed with the idea of picking up Ottawa's share of the project or bringing in other investors, like the Province of Ontario, to help finance the operation.[47] It quickly became apparent, however, that finding another investor willing to pony up the money for an expensive oil sands operation with dubious economics was difficult, if not impossible. After months of protracted negotiations, OSLO's project chairman, Gordon Wilmon, acknowledged publicly what few in the provincial government cared to admit. "It's going to be very difficult to proceed straight ahead without some delay. This simply may not be the right time."[48]

Gradually, the realization grew that Alberta's oil industry needed to become more competitive, particularly the unconventional sector. Canadian oil and gas began to lose its market share in Ontario to more inexpensive imports. In February 1990, Canada became a net importer of oil for the first time since 1982.[49] By the spring of 1990, the Province of Alberta, Petro-Canada, and Gulf Canada – representing more than 50 per cent of the stakes in the project – all announced their willingness to sell their shares in Syncrude.

Although Petro-Canada and Edmonton made it clear the move would free investment capital to put into other oil sands projects, cash-strapped Gulf Canada began to sell off all its heavy oil holdings. Combined with Ottawa's reconsideration of OSLO, the slow sale of Syncrude shares reflected investors' unease with the perceived uneconomic nature of oil sands operations.[50] Increased maintenance and upgrading costs combined with low world crude prices to prevent Syncrude from posting a profit for five straight years. Edmonton's willingness to forgo royalty payments after a costly fire at Syncrude encouraged perceptions that the oil sands' current economics were unwise.[51] Assessing the bleak situation in May 1990, *The Financial Post* concluded that "'heavy oil' is threatening to become a dirty word in Canada's oil patch."[52]

The province floated several suggestions to sustain OSLO for the next two years, but none managed to attract much interest. But in May 1992, Rick Orman visited the Middle East and toured Saudi Arabia's vast oil fields. The visit offered Alberta's Energy Minister a first-hand look at the sheer size of Middle Eastern reserves. The large producing fields, combined with Saudi Arabia's low export and production costs (around $2 per barrel), lead Orman to recognize that Alberta needed to become more competitive with international oil.[53]

Five months later, on October 15, 1992, the province announced the first significant changes to Alberta's oil and gas royalty system since 1973. The province introduced a permanently lower royalty rate for any new pools of oil discovered and concessions for existing or inactive wells. Taken together, the royalty cuts slashed $240 million from provincial revenues.[54] Although OSLO lingered for a few weeks after Orman's royalty rate announcement, the writing was on the wall. Finally, in early November, OSLO's partners announced the death of the project. "The days of the $4 billion to $5 billion project are probably over," said OSLO's Chairman, Gordon Wilmon.[55] Perhaps the only balm to soothe Alberta's wound came when Ottawa mothballed OSLO's chief rival for federal funds, Hibernia, in 1992. Along with the demise of the last of the great mega-projects came the realization that the great oil boom of the 1970s and early 1980s was over.

In retrospect, OSLO's collapse was a blessing in disguise. It employed virtually the same mining and separation methods used at both the Suncor and Syncrude plants and did not represent the significant technological step forward needed to be competitive. Instead, it relied on loan guarantees and tax and royalty concessions to be profitable. Throughout the search for additional investors after Ottawa's decision to pull out of the project, the provincial government tried to play the energy sufficiency card. Repeatedly, provincial government and industry spokespersons emphasized that diminishing conventional crude reserves made the development of the oil sands imperative.[56] Even the first Gulf War between the U.S.-led coalition and Saddam Hussein did little to save OSLO.[57] With abundant, and cheap, oil supplies available on the world market, few outside the province listened. To survive, the oil sands industry had to compete with conventional crude supplies on its own merits.

THE "ALBERTA ADVANTAGE"

While Orman's royalty announcement was significant for the conventional industry, the real catalyst for Alberta's change was the 1992 election of Ralph Klein.[58] Klein replaced Getty as leader and inherited a bloated bureaucracy and an angry electorate. With only six months to prepare for the next election, Klein moved quickly to distance himself from the failed policies of the Getty years. He reduced the size of the cabinet to seventeen ministries from Getty's twenty-six. Government radically cut $700 million from spending as the province committed itself to introducing a balanced budget by 1996–97. Furthermore, $2.1 billion in loan guarantees made by the former premier were lost in the first few months of Klein's tenure and another $12 billion at high risk because the companies the province supported were financially unstable.[59] The solution to the province's woes seemed less government, not more, and Edmonton announced plans to divest itself of shares in various industries. Government, said Klein, "had over-promised and under-delivered; it had tried to be a provider of things it should not provide and could not afford. We are saying that we understand our limits; that we can't provide some of the things that we used to; that we do not have all the answers."[60]

Provincial politics veered in a new direction that mimicked larger, global trends. Nevertheless, the moves baffled political commentators seeking to make sense of new issues through the prism of old paradigms. Klein never specifically stated that Alberta had embraced globalization once he assumed office; he only allowed that the "Alberta Advantage" of lower provincial taxes and investor-friendly climate stimulated economic growth.[61] "The Progressive Conservative party came to power in 1971 determined to make Alberta prosperous by making the government bigger," wrote Lisac who laboured hard to understand what was going on in the province. "That leadership reinvented itself in the early 1990s, and held on to power with a determination to make Alberta prosperous by making the government smaller."[62]

Students of globalization, however, saw no mystery behind many of Klein's actions. In fact, the new policies bore striking similarities to steps taken by other governments around the world since 1979 when British Prime Minister Margaret Thatcher launched the "Market Revolution."[63] The policy of reducing the role of the state, instituting lower taxes, privatizing state-owned companies, and returning initiative to the market became so synonymous with Thatcher that the movement soon bore her name. Thatcherism declared "big government" the enemy. "We should not expect the state to appear in the guise of an extravagant good fairy at every christening, a loquacious companion at every stage of life's journey, and the unknown mourner at every funeral."[64]

Thatcherism moved to the United States with the election of Ronald Reagan in 1980. The fortieth president pledged to restore America's international prestige after the trials and tribulations of the 1970s. Moreover, the "Reagan Revolution," challenged the postwar New Deal-Great Society consensus and argued that "big government," in the form of the modern welfare state, needed to be scaled back. "In this present crisis," stated Reagan during his inaugural address, "government is not the solution to our problem."[65] Instead, the new president argued, "if we cut tax rates and reduce the proportion of our national wealth that was taken by Washington, the economy would receive a stimulus that would bring down inflation, unemployment and interest rates."[66]

In his analysis of globalization, *The New York Times* foreign correspondent Thomas Friedman noted the increase in the number of countries that began adopting similar philosophies in the 1980s. The similarities between developments in Britain, the United States, and the Asian "Tigers" of Indonesia, Thailand, Malaysia, and Japan were striking. Friedman characterized these moves as the economic equivalent of putting on a "Golden Straightjacket" because the process limited government spending and made the private sector responsible for economic growth. The state privatized industries in communications and manufacturing and drastically reduced the size of the bureaucracy. Moreover, governments sought to keep inflation down, reduce tariffs on imported goods and open internal markets to international competition. "When you stitch all of these pieces together," wrote Friedman, "you have the Golden Straitjacket."[67]

To some extent, the free market revolution made tentative inroads into Canada by the early 1990s but no politician – federally or provincially – dared to embrace the philosophy wholeheartedly before Klein.[68] Nevertheless, the nationalist experiments and Keynesian policies associated with the Trudeau years failed to stimulate the economy. "Industrial strategies aimed at subsidizing 'winners' to make Canada globally competitive had not had any noticeable effect," wrote historian Michael Bliss.[69]

Structural changes to the Canadian economy also encouraged new directions for economic development. Shortly after the Canada-U.S. Free Trade Agreement went into effect, the United States and Mexico announced plans to pursue a trade and investment program. The Canadian government was aware of the negotiations and a Cabinet Task Force urged the government to participate in future discussions.[70] "If the perceived gains of the Canada-U.S. bilateral accord were to be defended," wrote historians John Thompson and Stephen Randall, no real alternative existed but "to make the negotiations trilateral."[71] Despite initial reservations about the desirability of another trade agreement, the Canadian government pursued the agreement until ratification by Jean Chrétien's Liberal government in January 1994.[72]

The North American Free Trade Agreement (NAFTA) did not significantly alter the energy agreements between Canada and the United States but reinforced Ottawa's willingness to pursue regional trade liberalization. Despite the Liberals' vociferous opposition to the FTA five years earlier, a Liberal Prime Minister ratified NAFTA with barely a second glance. The political *volte-face* reflected the extent to which globalization altered the parameters of public policy debate in Canada. "The Golden Straightjacket narrows the political and economic policy choices of those in power to relatively tight parameters," writes Thomas Friedman. "Once your country puts it on, its political choices get reduced to Pepsi or Coke."[73]

THE NATIONAL OIL SANDS TASK FORCE

The change in government philosophy and emphasis on free market enterprise presented a unique opportunity for the oil sands industry to serve as a focal point for Canada's economic resurgence. Furthermore, after the collapse of OSLO in 1992, the industry realized it had to change its approach to compete effectively for investment dollars. Industry would apply the "cold shower" of competition to trim the excess off bloated Canadian companies in the oil sands.

In March 1993, the Alberta Chamber of Resources established the National Oil Sands Task Force to bridge the gap between industry and government over oil sands development. Representatives gathered from all sectors of the oil industry and other interested parties, including the federal and provincial governments, the National Research Council, and the Canadian Imperial Bank of Commerce. Subtle changes in industry and government attitudes enabled the report to showcase government's new emphasis on creating a more favourable investment climate. Signals from government and industry were clear; Canada's synthetic fuels industry prepared to enter a new era of co-operation with government.

The industry did not regard the sands as a marginal resource operating on the fringes of the oil industry. By the early-to-mid-1990s many came to see Canada's synthetic fuels industry as a "knowledge-based, technology-driven, resource of substantial quality and value."[74] "The vision of the Task Force is to encourage additional economic investment in the oil sands so that the industry can triple production over the next 30 years," wrote the authors of the Task Force's final report.[75] For eighteen months, the Task Force held hearings between government and industry before producing its final report in 1995. The report outlined a comprehensive strategy for oil sands development into the 21st century that emphasized teamwork between industry and government. "Why do we have to compete with each other?" asked Chairman Erdal Yildirim. "There is no competition. You compete against Venezuela. You compete against Mexico. You compete against other parts of the world that produce heavy hydrocarbons."[76]

The Final Report identified eight "levers of development" that, if correctly manipulated, would result in optimal oil sand development in the next twenty-five to thirty years:

1. A market-driven science and technology innovation system that results in sustainable development and lower supply costs
2. New generic, competitive and fixed royalty and taxation regimes
3. Diverse, internationally based capital finance formation
4. Sustainable development and environmental compliance
5. Aggressive national and international marketing for bitumen and oil sands products
6. A complete pipeline transportation system to ship product to market
7. "Fair, predictable, timely and competitive regulation"
8. Informed and committed stakeholders[77]

Because the oil sands were such a high technology, knowledge-based industry, investment in science and technology would help reduce finding costs. Only by continuing to invest in research and development could the province expect to make the oil sands economically viable. The report pointed out that industry-sponsored research reduced operating costs from C$30 per barrel in 1985 to C$14 in 1995.[78]

New technologies developed through government and industry research helped reduce oil sands costs in both conventional mining and *in situ* projects.[79] Simple changes, like using large computerized trucks and shovels, radically altered the economics of mining operations.[80] Both Suncor and Syncrude relied on draglines and conveyor belts to mine the sands and transport them to the extraction facilities. This meant that the first two oil sands operations only had access to part of the deposit. Redeploying the system within the deposit proved costly and was subject to breakdowns in the winter months. The multitude of moving parts and machinery required to sustain operations meant that the two companies dreaded breakdowns, which could cause huge production bottlenecks. By changing to trucks and shovels, the two companies reduced operating costs and increased their profitability by reaching portions of the deposit that contained richer quantities of bitumen.[81]

The application of new technologies and improvements in horizontal drilling techniques breathed new life into *in situ* projects. Imperial Oil's Cold Lake project relied on Cyclic Steam Stimulation (CSS) to heat the deposit and collect the bitumen. CSS requires companies to pump steam into the deposit and then to wait one month for the moisture to soak into the deposit before pumping the bitumen to the surface. The difference between CSS and Steam Assisted Gravity Drainage (SAGD) is that SAGD is a continuous process that does not require long delays between operations. Drillers place two parallel horizontal wells, approximately five meters apart, into the deposit. In the first well, they inject superheated steam into the deposit. Heat "cooks" the oil sand and separates the oil and bitumen from the sand. The liquid then collects in the second well before it is pumped to the surface.[82]

Cold Lake witnessed the first complete SAGD test in 1978. Although the tests gave encouraging results, replicating the process elsewhere was difficult. Drillers could not guarantee producers that they could drill two parallel wells within five meters of one another, a requirement for the process to work. When horizontal drilling techniques improved in the late 1980s and early 1990s, they eliminated the final obstacles to the system. Computer-controlled drilling rigs accurately controlled the distance between the two wells. SAGD has many advantages over the CSS process, not the least of which is its lower operating cost. The introduction of SAGD was the single most important development in oil sands technology since Karl Clark's work on hot water separation. New technology enabled smaller companies to join with the majors in developing the sands.[83]

Despite these important developments, the report argued that the practice of funding several autonomous research projects wasted resources. The report concluded that many of these technological advances occurred in spite of the industry. In 1974 when the Province of Alberta created AOSTRA, approximately twenty people – perhaps ten academics and an equal number of their students – were conducting research on the oil sands. Twenty years later, more than eighty science and engineering professors were teaching, doing advanced research and development on oil sand-related projects.[84] While some inside the industry coordinated their efforts, others did not. Inevitably, the lack of coordination resulted in needless duplication as companies isolated their research teams from the competition. The task force suggested that industry would benefit if researchers used a "collaborative alliance" of resources to direct R&D funding "to the most important technology needs."[85]

The pace of technological developments would determine the future of oil sands developments; industry participants quickly fingered the current fiscal regime, or royalties and taxes, as the main obstacle. The final report reflected corporate concerns and specified the need for a universal tax and royalty regime to apply to all oil sands projects in its final report. The task force proposed that the province adopt a generic royalty system based on a percentage of net project revenue until recovery of all start-up costs. This, argued the task force, would accelerate the development of the oil sands while providing the Province of Alberta with a fair return. The universal regime would also eliminate uncertainty in the business community by giving companies reasonable "ground rules" for development. Developers could anticipate costs with a higher degree of certainty, encouraging other companies to develop deposits. Improved financial terms would also help oil sands development by attracting capital from a wide range of sources, including banks, insurance, mutual funds, and the equity markets.[86]

After the release of the final report, the province immediately began discussions on the task force's recommendations. On September 6, the

Standing Policy Committee of the Alberta government approved the generic oil sands regime. Two months later, Premier Ralph Klein announced that the new royalty regime applied to all new projects. "This is an example of the government's new approach to development," said Klein. "Instead of participating directly, we are establishing a framework that should encourage new projects, which mean more jobs and a stronger Alberta – and Canadian – economy."[87] Finally, after decades of industry lobbying, the province implemented a generic royalty and tax regime that would apply to all oil sands projects. The province would receive a minimum royalty of 1 per cent on all production. The royalty would increase to 25 per cent on net project revenues after the project developer recovered all start-up costs, including research and development costs and a return allowance. More important, for project developers, all capital costs – including operation, and research and development costs, would be 100 per cent deductible in the year incurred.[88]

The Alberta Department of Energy estimated that the province could expect a short-term decline in royalties because of the new regime. Existing projects – Syncrude, Suncor, and Cold Lake – could write off reinvestment and plant upgrades while lower world prices would reduce the money government received. Nevertheless, the Alberta government recognized that the new policy would pay dividends over the long term and would encourage new investment in the oil sands. "This royalty regime design is intended to provide an incentive for the capital investments necessary for the development of the oil sands," concluded the Department of Energy. "The bulk of royalties will be shifted to future years."[89] Combined with the federal government's decision to apply a universal tax regime to all oil sands projects, the industry finally had an overall development plan.[90]

Meanwhile, the federal government modified its taxation and royalty regime as well. Previously, Ottawa's tax scheme distinguished between conventional mining projects and *in situ* projects, favouring the former by allowing a wider range of deductions. The 1996 budget eliminated the distinction between the two as the federal government carried out a universal tax regime for oil sands producers. In addition, Ottawa extended new tax incentives to spur the industry. Before 1996, Ottawa provided incentives for project expansions but disallowed write-offs related to upgrades aimed at improving efficiency. The 1996 budget changed these provisions and helped streamline the policies of both the federal and provincial governments.[91]

The new strategy significantly stimulated oil sands development. In 1997, production from the sands increased to more than 540,000 barrels per day, an increase of 18.6 per cent over the previous year.[92] New taxation policies also triggered an investment boom as several small and medium-sized companies, like Koch Oil Sands, Murphy Oil and Black Rock Ventures, invested in the oil sands.[93] Technological developments, particularly the development of SAGD, meant that these smaller companies did not have to invest billions of dollars, like Imperial Oil, Petro-Canada, and Syncrude. Within three years of the task force's final report, the industry estimated that companies planned to invest more than $24 billion in oil sands projects. "Future production of synthetic crude oil from mining and *in situ* projects is anticipated to increase even more significantly," noted the EUB, "as refined products from the oil sands replace the depleting conventional oil and gas reserves of the province."[94]

Therefore, the agenda for the second generation of oil sands producers was complete by 1997. The new oil sands development strategy encouraged investment and expansion. National Energy Board assessments reflected the newfound optimism and projected that production from the oil sands would exceed 1.6 million barrels per day by 2015.[95] "An opportunity for Canadian producers to sell their increasing outputs of synthetic crude oil and blended bitumen is developing in the North American marketplace," concluded the NEB. Still, the NEB also warned that synthetic producers might have to offer American refineries a discount to accept greater amounts of oil sand oil.[96] The dilemma, according to the EUB, was that technological advances made the oil sands more profitable for producers but conventional crude sources still enjoyed a significant price advantage. Conventional crude, particularly from the Middle East and Latin America, operated with lower production or transportation costs than synthetic crude. Although the oil sands could expect to keep its stranglehold on the Midwest market, making gains elsewhere would be difficult.[97]

CONTINENTAL ENERGY AGAIN?

Several issues remained before the oil sands gained widespread acceptance as a reliable source of energy for North America. Chief among these were environmental concerns and the need to improve the competitiveness of the oil sands *vis-à-vis* cheaper sources of foreign crude. Many of these issues came into focus when the profile of the oil sands increased during the "new" energy crisis in early 2001. Rolling blackouts through California and sky-rocketing heating fuel costs threatened the prosperity of the United States. After President-elect George W. Bush came to office in January 2001, he asked Vice President Dick Cheney to put together a comprehensive energy strategy for America's future. As the vice-president focused on this task, the vast oil sands deposits in Northern Alberta came to his attention. Canadian Prime Minister Jean Chrétien encouraged the attention, and offered Alberta's oil sands as the answer to America's energy woes.[98]

Despite the prime minister's enthusiasm for expanding markets available to the oil sands, his positive message to President Bush obscured deep divisions within the Canadian government. At the federal level, politicians regarded the Bush administration's interest in a continental energy strategy with some degree of ambivalence. While they wanted the United States to look at the oil sands as a valuable fuel source, they were also leery of expanded trade with the United States.

Their ambivalence also reflected environmental concerns about the impact of greater development. Many environmentalists considered the oil sands to be a dirty source of energy that would unnecessarily prolong dependence on fossil fuels. Canadian environmentalists protested further expansion of oil sands projects and the debate on continental energy increasingly became presented as a choice between energy and the environment. On February 23, 2001, the David Suzuki Foundation strongly urged the Chrétien government to reject any continental energy pact. "If Canada agrees to boost its oil and gas production to meet U.S. demands, it will cause our greenhouse gas emissions to rise and worsen climate change," said Gerry Scott, the Director of the Foundation's Climate Change Campaign. "The emissions from the production of a 150,000-barrel-a-day oil sands operation is equal to the greenhouse gas emissions of 1.35 million cars."[99]

Chrétien quickly retreated from his earlier statement after recognizing that the proposed expansion of oil sands development would conflict with his environmental agenda.[100] In a nod to Canada's environmental lobby, the government reaffirmed its support for the Kyoto protocol in late March and again in early June 2001.[101] The statements seemed to erode Chrétien's initial offer of increased oil sands production to offset supply shortages. "We have an obligation to show strong, visible leadership in reducing emissions," said Energy Minister Ralph Goodale. "To reduce energy consumption, we will make energy efficiency improvements on buildings, put the federal garage in order, and buy more 'green power.'"[102]

Before the announcement of the Cheney plan, Canadian politicians made it widely known that they expected the United States to address several issues, including conservation. "We need [to see] an emphasis on energy efficiency and energy conservation," said Energy Minister Goodale before the Standing Committee on Aboriginal Affairs, Northern Development and Natural Resources. "These things are fundamental characteristics of an intelligent society, and they need to be given appropriate attention along with the supply issues."[103] During his testimony, Goodale emphasized that he did not intend to pursue a common North American energy policy with the United States. "Canada has and will continue to have its own energy policies based upon open competitive markets, fair, and efficient regulation, and the principles of sustainable development. But we do believe we can expand North American energy markets to the advantage of both consumers and producers."[104]

The federal government also had to contend with the suspicions of certain members of the oil industry and the Opposition regarding continental energy. Would the voracious appetite of the United States for energy translate into an energy shortage for Canadians? The discussion about a continental energy strategy was particularly disconcerting to the New Democratic Party who wanted to return to the days of the NEP instead.[105] The Chrétien government approached the issues very cautiously and carefully chose its words in public statements and debates. When asked if Canadians faced an energy crisis similar to the one that gripped the United States, Energy Minister Goodale answered that Canada did not. "We do not have a crisis but we do have an enormous opportunity .... We are not pursuing a North American energy policy. We are pursuing the expansion and successful functioning of energy markets in the Canadian interest."[106]

Although the distinction may have seemed clear to members of the government, it was less clear to many members of the House of Commons and the Senate. In Opposition circles, suspicions grew that a continental energy strategy would circumvent market forces. Canada would be forced to sell energy exclusively to the United States at the expense of meeting Canadian demand.[107] In April, when the prime minister created an energy cabinet committee and closed off outside scrutiny of decisions and debates, Opposition cries became more intense. "Since the Americans already have the right of access to Canada's energy supplies and we have the right of access to the U.S. markets in oil, gas and hydro under agreements negotiated by the Conservative [Mulroney] government, what, exactly, is there left to share?" asked Progressive Conservative Senator Pat Carney.[108]

Questions also abounded about the long-term role of the oil sands in a continental energy policy when U.S. energy policy shifted frequently as administrations changed. The perception was that traditional problems remained: synthetic fuels were capital intensive, required long lead-times, and were environmentally unfriendly. Overriding these concerns was a pragmatic assessment that if Canada launched another "crash" program to develop the oil sands, the direction of U.S. energy policy might shift before completion. There was no guarantee that a different office holder would not overthrow a continental energy policy – no matter how vigorously pursued by President Bush. George W. Bush, opined one MP, "will probably only be [in office] for four years. Then what will we do if we have developed the tar sands at that point?" Although the government dismissed the question as simplistic – the United States requires energy whether a Republican or Democrat occupies the White House – the question pointed out the need for long-term planning and continuity in the development of synthetic fuels policy.[109]

Given the important role occupied by the oil sands in Canada's energy future, perhaps the point of some of these questions was moot: Canada still required a significant amount of petroleum for the foreseeable future. Public and private sector participants have invested in the oil sands for the last eighty years. Even before Cheney began his investigation, industry sources alone invested between $30 and $35 billion in various oil sands-related projects. Quite simply, synthetic fuel projects would not disappear overnight. Some Opposition members bemoaned the lack of investment in renewable energy sources and claimed that North America should be following Europe's example and invest more in renewable energy sources.

Others dismissed the argument as impractical. The development of alternative energy sources (wind power, solar power, and nuclear) would be unable to replace all the important services currently provided by hydrocarbons. "Green Germany," countered Alliance MP Bob Mills, "run by a green government is using nuclear energy for 70% of its energy needs. We should not believe the myth about Europe being a wonderful example to follow because it is just not there."[110] Where the government had a decision to make, however, was on the scale of development for the oil sands. Policymakers would have to consider the trade-off between expanded energy supplies and increased emissions from the oil sands.

As the debate about Canada's role in a continental energy strategy intensified, the Senate's Standing Committee on Energy, the Environment and Natural Resources held hearings to gauge public attitudes. The broad scope of the investigation ensured that the federal government heard from a wide range of representatives from different interest groups. These included representatives from the David Suzuki Foundation, advocates of solar, wind, hydro, and nuclear energy, the Canadian Association of Petroleum Producers, and cabinet officials. Through sustained investment, the Canadian government estimated that technological advances reduced the environmental footprint of a typical oil sands operation by approximately 25 per cent. Still, Energy Minister Ralph Goodale acknowledged room for improvement remained. "One of the technologies that should be examined with great care," said Goodale, "is how to capture the CO_2 right at source, before it is emitted into the atmosphere, and find a way to sequester it underground. If that technology with respect to a facility like the oil sands can be perfected – and I believe it can be – we'll take an enormous economic, environmental, and technological step forward."[111]

The Canadian Association of Petroleum Producers (CAPP) urged the Canadian government to continue cautiously.[112] In early April 2001, the organization prepared a briefing paper distributed to both the Canadian and American governments outlining the organization's approach to discussions about a continental energy strategy. Instead of advocating a purely continental approach, CAPP proposed that both governments continue to build on the success of the North American Free Trade Agreement and the Canada/United States Free Trade Agreement. CAPP's vice-president Greg Stringham agreed that a continental energy policy would make some "incremental improvements" to the system, but maintained that the creation of a continental energy policy was unnecessary. As CAPP pointed out in its briefing material, "We know the consequences of policies, however well intentioned, that restrain and prevent markets from operating freely and competitively. The cost to society is huge and far outweighs any perceived short-term benefit."[113]

After the Bush administration's energy policy announcement of May 2001, international events continued to frustrate attempts to push a more comprehensive energy package through Congress. World crude prices dipped below $25 per barrel and gasoline prices dipped about twenty cents per gallon. Because of a relatively mild summer, California managed to avoid a real energy crisis and managed to post an energy surplus. With public urgency about an energy crisis fading, the Bush plan seemed dead in the water. However, as Abraham McLaughlin pointed out in *The Christian Science Monitor*, President Bush possessed several options. "Of the 105 recommendations in the original plan, 89 can be implemented by presidential fiat through federal agencies or executive orders. Only 20 need to pass Congress." While the administration endured a public bruising in the debate over drilling in Alaska's Arctic National Wildlife Refuge and off the Florida coast, the media paid less attention to the administration's bilateral negotiations with Mexico and Canada. "In truth, many of the things that we would [hope for] have more to do with the passage of regulations," said Edward Murphy of the American Petroleum Institute.[114]

Through the summer of 2001, Bush's energy strategy began to falter even as Alberta Premier Ralph Klein travelled to Washington to meet with energy officials. "We have so much energy to burn, so to speak," said Klein after his meeting with Cheney, "and we're willing to share."[115] Although Alberta's premier was confident that the oil sands would play a major role in North America's energy future, rumblings emerged from the Prime

Minister's Office that federal officials did not concur. Nor did the Prime Minister endorse the intent of Klein's trip to Washington. The federal government maintained that it, not the Province of Alberta, determined trade policy with the United States.[116]

The renewed bickering between Edmonton and Ottawa clouded the fact that the international community began to focus on the significance of the oil sands. In the late summer, Senator Nick Taylor, Chair of the Canadian Senate's Energy Committee, toured several European countries on a fact-finding mission about nuclear energy. Upon its arrival in Vienna, his delegation decided to pay a courtesy call on OPEC. Previously, no formal contact between OPEC and the government of Canada had ever taken place. On this day, for whatever reason, OPEC agreed to a thirty-minute courtesy call to discuss the use of nuclear energy. However, when Taylor and the rest of the delegation arrived in the ministry's office, the informal gathering soon evolved into a high-level policy meeting.[117] OPEC ministers expressed a great deal of interest in the Alberta oil sands. While they made it clear that they did not feel threatened by the emergence of a significant oil sand industry in Canada – Canada remains a regional, rather than global, supplier of crude oil – they asked probing questions about the project's environmental safeguards. Did the high sulphur content pose a particular danger to the environment, particularly if the refining and upgrading of bitumen took place in such a confined area?

The most intriguing development took place when OPEC floated a test balloon about Canada joining the cartel. With Canadian production approaching one million barrels per day, it made sense that OPEC would sound out Canadian policymakers. Canadian membership in OPEC would avoid the cartel's biggest fear about a continental energy strategy – that North America would become energy self-sufficient and would build a large tariff wall around the continent, denying OPEC access to the market of its single largest customer, the United States.[118]

Bush administration suggestions in early 2001 that the United States should reduce its reliance on imported Middle Eastern oil aroused concerns in Vienna.[119] Moreover, the move toward a continental energy strategy began at roughly the same time that OPEC nations began preliminary development of Middle Eastern natural gas supplies. Natural gas is the fuel of choice for many electricity generating facilities. Demand for gas by electric companies in the winter of 2000/2001 meant that electricity prices served as the determiner of natural gas prices. High electricity prices meant that electricity-generators could afford to pay high prices for natural gas.

However, because of its volatile nature, continental energy sources typically supplied natural gas to the United States (imported as liquefied natural gas – LNG). The United States consumes all the natural gas it produces and Canada supplies the difference. Between 1985 and 2001, Canadian natural gas exports to the United States quadrupled and met approximately 15 per cent of U.S. needs. This sum represented approximately $50 billion in export revenues and more than 50 per cent of Canada's trade balance. Thus, as prices for natural gas reached more than $4.00/mcf (thousand cubic feet), American companies began investigating the feasibility of importing natural gas from abroad. OPEC also saw an opportunity to market a new product. As the world's third-largest producer of natural gas, OPEC has a stake in increasing its share of the energy market through alternative sources.[120]

September 11, 2001, brought the issue of continental energy back into focus, particularly since many terrorists responsible for the attack on the World Trade Center in New York and the Pentagon in Washington, D.C., had ties to Saudi Arabia. The attacks, said former Clinton Energy Secretary Bill Richardson, "should be an alarm bell that we need a balanced, comprehensive energy policy that addresses things we don't like to do: mandating more fuel-efficient vehicles, more domestic oil and gas drilling, becoming more energy efficient as a nation."[121] The Department of Energy claimed that the National Energy Plan provided a suitable blueprint, particularly in reducing imports from the Middle East and relying on continental energy. Still, the administration encountered resistance from environmental groups who opposed drilling in Alaska and upgrading oil sands facilities in Northern Alberta. "These days, everybody wants to be considered an environmentalist," said Robert Kripowicz, the Department of Energy's Assistant Secretary. "Opponents make the point that they are environmentalists, but I believe those of us who support environmentally responsible energy development can make the same point."[122]

Saudi Arabia's tepid support of the war on terrorism opened the door to questions about the political and military costs of relying on Saudi Arabian crude. "The stark truth is that we're dependent on this country that directly or indirectly finances people who are a direct threat to you and me as individuals," said Edward Morse, former deputy assistant secretary of state during the Reagan administration. "They won't give us information, won't help track people down, and won't let us use our bases that are there to protect them."[123] Although the forecasted rupture in U.S.-Saudi Arabian relations did not materialize, the relationship clearly endured a difficult storm.[124]

Other issues affecting the future of oil sands development remained, particularly environmental concerns. Canada's commitment to the Kyoto treaty loomed as the single greatest threat to the future of the oil sands because of the uncertainty it created for investors and stakeholders in oil sands projects. Who would bear the costs of complying with Kyoto? Would the federal government push ahead with ratification before providing Canadians with a detailed blueprint? Ratifying Kyoto, pointed out Alberta Energy Company President Gwyn Morgan, could result in the increase of atmospheric CO_2 by penalizing efficient countries, like Canada. Kyoto's detractors point out that treaty nations only comprise 30 per cent of the world's CO_2 production. Instead of rewarding countries that have attempted to improve efficiency, development opportunities would go to "heavy polluters," like China. Lacking competition, these countries would race ahead with development, whatever the impact on the environment.[125] In part because Kyoto excluded developing nations from making emissions reductions, President Bush withdrew U.S. support for the treaty in March 2001.

Jean Chrétien remained personally committed to the Kyoto protocol through the first quarter of 2002, but said that Canada would seek "emissions credits" for cleaner burning fuels. "If we're shipping clean natural gas down there [to the U.S.] so they can replace coal, it reduces the world's CO_2," said Alberta Environment Minister Lorne Taylor. "Why shouldn't we get credit for it?"[126] In April 2002, the European Union's Environment Ministers balked at the plan and accused Canada of dragging its feet on the issue.[127]

Despite howls of protest from environmentalists, several factors argued against immediate ratification of Kyoto. Through the first quarter of 2002, fears grew that if Canada ratified the agreement in its present form, the future of the oil sands development would be jeopardized, forcing the country to increase its reliance on imported crude. Other industries sounded caution bells as well. The Canadian Chamber of Commerce estimated that enforcing Kyoto would cost the Canadian economy approximately $30 billion.[128] The Canadian Manufacturers and Exporters agreed with the Chamber of Commerce and concluded implementation would cause the permanent loss of 450,000 jobs in the manufacturing sector alone.[129] Chrétien's announcement on April 15 that Canada would ratify Kyoto "one day" elicited relief throughout the business sector. But instead of shelving the project indefinitely, the federal government was biding its time to get its house in order.

CHAPTER 9

Green Patch? Oil Sands Development in Post-Kyoto Canada

Ratification of Kyoto is one thing, the rules for how Canada
would allocate the burden of meeting Kyoto is something
else. Would allocation be by region, or would it be by sector?
We simply don't know at this point. – *David May, Alberta
Environment Spokesman, December 2001*

On September 2, 2002, Prime Minister Chrétien surprised opponents and
delighted environmentalists by announcing that his government intended
to ratify the Kyoto protocol before the end of 2002.[1] Chrétien's announcement forced the industry on the defensive and cast a pall of uncertainty
over the future of Canada's oil sands development and the security of North
American energy supplies. According to the terms of the agreement signed
at Kyoto, Canada pledged to cut its greenhouse gas (GHG) emissions in
2012 by 6 per cent – to approximately 570 megatonnes.[2] Estimates project that, by 2010, Canada will produce approximately 809 megatonnes of
greenhouse gases – roughly 240 megatonnes over the Kyoto-imposed limit.
For a country with a growing economy and expanding population, the task
ahead is daunting.

If perception is everything in politics, the Kyoto Accord represents
a striking dilemma for Canadian politicians and business leaders alike.
On the one hand, many Canadians believe that something must be done
about the environment. Action on Kyoto therefore offers the Chrétien

government the opportunity to show some tangible results on addressing environmental concerns.

On the other hand, independent assessments from several different sources concluded that, even with Kyoto's enforcement, the accord's net effect on global warming will be negligible. The International Energy Agency (IEA) pointed out that, even with ratification and enforcement of Kyoto, worldwide greenhouse gas emissions will *increase* 70 per cent over the next two decades.[3] Significantly, the IEA suggests that nearly two-thirds of the emissions increase will occur in Annex II countries, most notably China and India, whose emissions are not limited by Kyoto. In fact, if current consumption trends hold, the IEA predicts that by 2020, China will be the second largest energy consumer in the world next to North America.[4] A report by the C.D. Howe Institute in late 2002 reached similar conclusions. Kyoto, according to the Institute's report could only delay future environmental damage; it could not stop it entirely.[5] Even if Canada were to ratify the treaty, Canadians account for such a small percentage (3.3 per cent at 1990 levels) of the world's greenhouse gasses that one European observer dismissed Canada's participation, or lack thereof, as irrelevant.[6] However, the worldwide Kyoto push stumbled after both the United States and Australia announced they would not ratify the deal. In the aftermath of these announcements, Canada's 3 per cent suddenly became very important. To take force, the Kyoto protocol had to be signed by fifty-five countries that accounted for 55 per cent of the industrialized world's GHG's in 1990.[7] Heading into the summit at Johannesburg, the requisite number of signatories had already been collected; however, those countries were responsible for 37.1 per cent of GHG emissions. Russia's pledge to ratify, representing an additional 17.4 per cent, would nudge the total to 54.5 per cent. "For a Prime Minister at the end of his mandate who is casting about for a legacy," wrote *The Globe and Mail*, pledging Canada to Kyoto and nudging the treaty into effect would "be a world-class coup."[8]

Aside from Canada's part in ensuring that the treaty would become international law several environmental groups, and the European Union, believed Canadian ratification was crucial. Without Chrétien's pledge, no countries in North America would follow binding emissions targets.[9] "Europe," noted Gerald Doucet, secretary general of the World Energy

Council, "has put tremendous pressure on Canada, in particular, to ratify the Kyoto protocol so that it would break this North American cabal." [10] A few weeks before Chrétien's Johannesburg announcement, repeated attempts to kick-start Canada-EU trade negotiations proved unsuccessful. [11] Although they have never linked the two issues before, the EU's tougher stance at international trade talks in 2002 suggested that some linkage may exist. [12]

For Canada's oil sands industry, Kyoto represented a leap into the dark. Initial forecasts pegged the cost of implementing Kyoto anywhere between $0.50 per barrel and $7.00 per barrel, the upper end of which would destroy producers' profit margins. [13] Assessing Kyoto's cost for industry was difficult because the extraction techniques varied. For example, while Suncor announced its assessment that Kyoto would increase production costs between twenty and twenty-seven cents per barrel, another company, Canadian Natural Resources Limited (CNRL) estimated costs would increase by seventeen cents per barrel. CNRL's projected costs were lower because their facility uses newer technology than Suncor. Still another company, Nexen, released figures showing increased production costs between thirty and forty cents per barrel. [14]

Although the initial per barrel costs for implementing Kyoto seemed relatively modest, the problem for synthetic producers lay in the nature of the industry, with its long lead-times and its need for large amounts of capital. Even with advances in oil sands recovery technology over the past two decades, the size and scale of oil sands projects still required companies to entice start-up capital from their investors between five and seven years before production began. Furthermore, costs for synthetic projects remained dependent on several different variables, including start-up costs, access to skilled labour, energy and diluent supply, repairs and maintenance to facilities, upgrading costs, environmental protection, and continuing research and development. However, the ultimate dilemma for the oil sands industry was that it remained an international price taker dependent on world crude prices. Additional costs, like the extra C$1.4 billion Suncor required to complete its C$2 billion Millennium Project expansion, are not reflected in final commodity prices. [15] Companies can only charge world prices and cannot, therefore, pass on additional costs to consumers.

Canadian oil sand projects compete against global competitors to attract investors and market share. Investors must consider a wide range of factors, including project viability, return on investment, and overall attractiveness before committing their capital. Notwithstanding the inherent risks in resource exploration, Canada's taxation regime did little to buttress investment. Currently, Canadian governments tax the natural resource industry at a higher rate (42–46 per cent, depending on the province) than other domestic industries (37–41 per cent) while global tax rates are much lower (approximately 30 per cent).[16] Oil companies feared that the additional expenses associated with enforcing Kyoto could result in investment dollars being spent outside the country. One energy company involved in oil sands operations, Nexen, began publicly pursuing development projects in the North Sea and West Africa in the midst of the Kyoto debate. Both international projects offered a potential return of 20 and 25 per cent while Nexen's oil sands venture promised a pre-Kyoto return rate of 15 per cent.[17] The longer Ottawa remained silent about the cost to producers of carrying out Kyoto, the more uncertainty plagued investors and industry. Without a concrete implementation plan, the Kyoto "debate" amounted to little more than a shouting match with little substantive content.

As rhetorical debates about the merits of Kyoto intensified after Chrétien's announcement, business leaders and political commentators, particularly in Alberta, drew comparisons between Kyoto and the reviled National Energy Program. Kyoto's critics alleged that the treaty amounted to little more than a second attempt to plunder Alberta's wealth for the benefit of Eastern Canada. Other comparisons between Kyoto and the NEP are striking. In each instance, public debate on the issue polarized Canadians along regional lines, pitting one province against another. In one plan put forward by a Quebec MP in early 2003, Alberta would be responsible for meeting 40 per cent of Canada's GHG reductions target; Quebec a mere 5 per cent. The rationale was straightforward: Alberta's oil and gas industry was "dirty" and emitted more greenhouse gasses than Quebec's hydroelectricity plants.[18] Battles for allocating greenhouse gas reductions expanded to individual industries. Eyebrows rose in early January 2003 when Ontario's auto industry received an exemption from meeting Kyoto's requirements. Instead, Ottawa asked automakers to improve fuel efficiency by 25 per cent by 2010.[19] Cynics in the oil patch were unconvinced by arguments that Ottawa could achieve better reductions through improving fuel efficiency in the end product and retorted that the prime minister would do anything to avoid harming Ontario's economy considering the number of voters in that province.

Comparisons between Kyoto and the NEP did not point out that, in both instances, the position of Canada's oil and gas industry enabled the federal government to isolate and marginalize the industry. In both 1980 and 2002, Ottawa could represent the oil industry's position as inimical to the national interest and priorities. Although five years passed between the government's initial agreement to Kyoto's provisions and its ratification of the accord, the industry failed during that time to mobilize significant public and political support to protect its interests. The debate also signalled the shifting landscape of energy policy in Canadian politics. In the late 1970s and early 1980s, the federal government framed the energy issue as one of energy revenues and self-sufficiency; environmental interests pushed industrial enterprises away from coal to cleaner-burning fuels like oil and natural gas. In the new millennium, environmental protection focused on reducing society's reliance on fossil fuels entirely. In each instance, both industry and the government of Alberta failed to articulate alternative strategies to deal with the federal government's agenda and win the critical battle for public opinion. Industry was unwilling, or perhaps unable, to assume a leadership role – to educate the public about the deleterious effects Kyoto could have on the economy; it failed to point out that government and industry already co-operated on reducing greenhouse gas emissions, and that improved technology had reduced the environmental footprint of individual oil projects and drastically reduced CO_2 emissions in oil sands and heavy oil projects. Industry paralysis forfeited the initiative to Ottawa.

In a textbook example of issue management reminiscent of the communications strategy employed by Marc Lalonde in 1980, the Liberals determined the direction of the policy agenda and marginalized opponents. The federal government repeatedly cited polls showing most Canadians, including Albertans, favoured ratifying Kyoto. In parliamentary debates, the Prime Minister and his government dismissed Opposition claims of an impending "investment chill" associated with ratification as nothing more than industry-sponsored scare-mongering. "We have heard the threat of investment loss before with the acid rain program and with the removal of lead from gasoline," noted the Chair of the House of Commons Environment Committee, Liberal MP Charles Caccia, in one typical exchange.[20]

By spinning Kyoto as the cornerstone of an environmental program – and not an energy package – Ottawa assumed a leadership role early in the debate it would never surrender. Kyoto's supporters hailed the decision as an important "first step" to arresting global warming and climate change. Others applauded the Prime Minister's pledge as courageous, particularly since the United States would not ratify the treaty. The decision, they argued, invoked the best traditions of Canadian domestic and foreign policy. Canada would make difficult choices and implement farsighted policies – not because they were easy, but because they were "right."

Unable to critique a concrete plan, and tactically outmanoeuvred by the government at every turn, the oil industry faced an uphill battle and found itself debating minor points instead. Some pointed out flaws in the claims about Kyoto's ability to improve air quality after the David Suzuki Foundation circulated a petition to Canadian physicians demanding the government ratify Kyoto to help stem air pollution.[21] Others pointed to opinion polls that suggested popular support for Kyoto was broad but shallow – polls consistently found that the more people learned about Kyoto, the less likely they were to support ratification.

Much to the industry's dismay, however, the people polled displayed a remarkable ignorance about the importance of hydrocarbon energy to Canada's economy and to their own lifestyles. One survey conducted for Natural Resources Canada found that 78 per cent of respondents thought that hydroelectric generators furnished the bulk of their energy. In reality, only 7 per cent of Canada's energy comes from hydroelectricity while 83 per cent comes from hydrocarbons (oil, gas, and coal).[22] What is important to observe, however, are the assumptions of the survey's respondents. Global warming could be solved by directing a national effort against a single, dirty, industry – oil and gas. Instead of examining personal consumption habits, the initial focus of the government enabled Canadians to believe that individual costs would be minimal. After all, a majority believed the bulk of their energy came from hydroelectric generators not subject to Kyoto's provisions.

Surveying the nature of the debate sobered objective industry observers. Not only did Canadians discount the importance of hydrocarbons for the country's future, the industry's reputation as a responsible corporate citizen emerged from the debate in worse shape than before. Afterwards, one oil patch analyst ruefully concluded that only the tobacco industry enjoyed a worse reputation than big oil.

Beginning from this strong position, the government's momentum hardly faltered once debate began in the House of Commons. To a degree, signals that Ottawa intended to ratify the treaty without a comprehensive plan caused some public discomfort as supporters and detractors argued over the costs of implementation. Skilfully, however, Prime Minister Chrétien quickly dismissed concerns by likening Kyoto's implementation to his government's 1993 strategy for retiring the federal deficit. "We didn't have a precise plan," explained Chrétien. "The government took a series of decisions. This is exactly the same thing." [23] The decision proved brilliant from Ottawa's point of view. The move increased the treaty's chances at ratification by denying Kyoto's critics something tangible to oppose and enabled the federal government to retain flexibility. When faced with uncomfortable questions, Prime Minister Chrétien declared that the government could make adjustments along the way. Canada, he added, could take as long as ten years to carry out a plan. [24]

Corporate officers throughout the oil patch, particularly those with large stakes in oil sands operations, remained unconvinced. They warned that uncertainty over Kyoto might force a re-evaluation of industry's commitment to the oil sands. One company involved in the oil sands, TrueNorth, a subsidiary of Koch Industries, Inc. of Kansas, announced that uncertainties over Kyoto, amongst other factors, prompted the company to trim $5.1 million from its fourth-quarter budget. Federal Environmental Minister David Anderson questioned the timing of the announcement. "They need better scriptwriters.... This is pretty clumsy stuff." [25]

Anderson's dismissive treatment of business concerns waved a red flag to the oil industry. Corporate officers viewed Kyoto's ratification as an abrupt ideological shift in business/government relations at the highest levels of Canadian government. Rather than continuing the "downsizing" trend of western governments, broadly attributed to globalization, the move smacked more of a return to the left-leaning, big government intervention approach employed by Pierre Trudeau.

Ottawa's own "Discussion Paper on Canada's Contribution to Addressing Climate Change" contained no less than five pages of "recommendations" for attaining greenhouse gas reductions in several industries. Recommending basic energy conservation measures (i.e., turning down the thermostat and installing better home insulation), required little government intervention. Other recommendations, like taxing petroleum products, converting existing highway systems to toll-roads, imposing a "sin tax" on Sports Utility Vehicles, increasing parking fees, and creating the Domestic Emissions Trading system (DET), would require the creation and maintenance of a pervasive, all-encompassing bureaucratic apparatus. Almost inevitably, given the size of emissions reductions Canada committed to under Kyoto – eliminating 240 megatonnes – enforcing the protocol would require massive government intervention in the economy. Equally likely is an emerging legal battleground of jurisdictional disputes and constitutional challenges.[26]

While no one would deny the fact that the various levels of Canadian government have the right and responsibility to establish a regulatory framework, government must also recognize that industry retains the right to choose where and when it will spend investment capital. If the burden becomes too onerous, the private sector is under no obligation to continue investing in Canada. As EnCana's Chief Executive Officer Gwyn Morgan wrote to the Prime Minister, implementing harsh policies would force businesses from all sectors of the economy to re-evaluate their commitment to Canada. "Any business wanting to grow," argued Morgan when discussing the options for companies seeking to purchase greenhouse gas emissions credits, "would have to pay a fee." In a global economy, the effect of such a decision could have devastating consequences on Canada's oil and gas industry and Canadian's energy future.[27]

A few weeks later, Morgan expanded his remarks in a speech before the Canadian Chamber of Commerce in London, Ontario, alluding to wavering investor confidence in the Chrétien government. Morgan pointed out that multinational corporations have no need to play politics in the manner suggested by Environment Minister Anderson the week before. In fact, Morgan concluded that businesses would "very unemotionally, very apolitically, say yes or no to Canada. I predict we will hear a lot more of these decisions. But there will be many more we won't even hear about, as investors silently move their money elsewhere."[28]

Morgan's open letter to the prime minister received no direct reply. Faced with Ottawa's stony silence on details, the provinces also began to stake out their positions. These developments were understandable considering that Kyoto's implementation also carried tremendous implications for the nation's provincial governments, particularly regarding allocating responsibilities for reducing Canada's greenhouse gas emissions. Only two provinces, Manitoba and Quebec, favoured quick ratification, while the others opposed the treaty to varying degrees. Ontario Premier Ernie Eves declared that his province would not support Kyoto unless the federal government explained how it intended to carry out the treaty. "I'm not signing on to anything that I don't know the effect of at the end of the day," said Eves. "It is incumbent upon the federal government to explain to every premier across the country ... [Jean Chrétien's] plan."[29] Saskatchewan Premier Lorne Calvert echoed Eves by indicating his province would oppose Kyoto unless Ottawa provided a detailed plan. "We've said we're willing to support the principles on which Kyoto is based," said Saskatchewan Industries and Resource Minister Eldon Lautermilch, "but we're not going to accept an unfair circumstance for our province."[30]

Despite the importance of Ontario's economy, and the echoes of support from Saskatchewan, no other province stood to lose more from Kyoto's ratification than Alberta. Perhaps more unsettling to both the province and the oil industry was the fact that the federal government sent mixed messages about what it intended Kyoto to accomplish. Did Ottawa intend to use Kyoto to make the Canadian economy more energy efficient or would Kyoto become the club to shut down the heart of Alberta's oil industry? Repeatedly, industry spokesmen emphasized their willingness to work with Ottawa but left frustrated by the lack of a concrete plan. Bitter experience and long memories conditioned many to expect that Ottawa would hammer the oil industry and spare others. "The production of oil and gas in

Canada makes up about 20 per cent of the carbon dioxide emissions," noted Greg Stringham in testimony before a House of Commons Committee in late November 2002. "The other 80 per cent comes from the consumption side, and that dialogue has not even begun yet."[31]

For CAPP and synthetic producers, the problem with Kyoto lay not in the requirement to improve efficiency and reduce greenhouse gas emissions; indeed, the history of the industry demonstrated that synthetic producers have actively sought to improve extraction, upgrading, and separation methods. Rather, the chief problem with Kyoto lay in the uncertainty it created. Would federal quotas on CO_2 emissions amount to a *de facto* cap on oil sands development? Would the royalty and tax regime change? What would happen to production costs if Ottawa required the oil industry to engage in international emissions credit trading on an open market? Would consumers willingly pay higher prices to purchase hydrocarbon energy? "The philosophy we've been coming to the table with is that if you put in requirements to reduce emissions to something below what is technically possible to do today, it doesn't matter what attitude [the industry has]," noted Stringham.[32]

In the November 18, 2002 CAPP position paper, CAPP expressed the industry's desire to work with government on global warming, but noted that the task ahead would be difficult. Although oil sands producers reduced their CO_2 emissions significantly, it remained inevitable that industry's overall emissions would continue to rise because domestic demand would increase. Declining conventional reserves make it essential that Alberta continue developing the oil sands if the country intends to remain a producer and supplier. Ottawa's own estimates suggest that oil sands production will exceed conventional crude production by 2005. Prior to Chrétien's Kyoto announcement, the National Energy Board projected that in 2015, Canada's conventional crude would supply between 15 and 20 per cent of Canada's energy needs. Oil from the sands would furnish between 66 and 75 per cent.[33]

Furthermore, in the aftermath of September 11, the role of Alberta's oil sands as a secure and dependable energy source for the United States expanded. In fact, through the first ten months of 2002, the average of Canadian crude exports rose by 1.5 million barrels per month.[34] Since refining and upgrading synthetic fuel emits slightly more greenhouse gasses than conventional crude (the 1995 Environmental Report published by the National Oil Sands Task Force revealed that the oil sands emit approximately 10–15 per cent more GHG than conventional crude. It was expected that total would drop to 8–10 per cent by 2000), the need to increase synthetic production to replace Canada's waning conventional supplies is at odds with the government's pledge to reduce Canada's emissions by 240 million tones.[35]

On December 18, 2002, federal Natural Resources Minister Herb Dhaliwal signalled that the federal government understood the industry's need for reassurances in the aftermath of Kyoto's ratification, particularly about Kyoto's potential costs. Dhaliwal announced that Ottawa would cap the oil industry's costs for purchasing CO_2 credits at $15 per tonne – or roughly $1.425 per barrel. In the same letter, Dhaliwal pledged that emissions intensity targets for the oil and gas sector would be set "at a level not more than 15 per cent below projected business-as-usual levels for 2010." Cautiously optimistic about the signals from Ottawa, EnCana's Gwyn Morgan called the move "a first step in the direction of maybe building some trust" back into the industry's relationship with the federal government.[36]

Ultimately, trust determined the course of developments in the Canadian oil sands. While it is possible that ratifying the Kyoto treaty could cost the oil and gas industry – and the Canadian economy – billions of dollars and thousands of jobs, the reality is that the total impact of the Kyoto treaty will not be known for decades. Nevertheless, the history of the oil sands suggests that if Canadian businesses, especially some crucial synthetic fuel producers, *believe* that Kyoto will put them at some competitive disadvantage *vis-à-vis* their competition, the industry could delay oil sands development indefinitely. EnCana, Canadian Natural Resources, and Petro-Canada – three major players in the oil sands – have all suggested their willingness to move part of their operations to the United States if Kyoto's application appeared contrary to their interests. This, in turn, could have other unintended consequences because Canada could lose the ability to control its own energy future. Without further expansion of the sands, the country's oil industry will contract and production will drop sharply, forcing the country to rely on increased imports to meet its needs.

AFTERWORD

I originally began work on this manuscript as a graduate student at Ohio University in 1999. For the completed dissertation, presented to the Faculty of Arts and Sciences in August 2002, mentioned the potential of Canada's ratification of the Kyoto Protocol before the domestic political debate really heated up. After the defence, I followed developments relating to the oil sands quite closely, talking to some of the sources I had interviewed during the course of my research in an attempt to keep the text as up to date as possible. The dilemma of writing about a current topic meant that revisions were required almost daily, particularly as the debate around Kyoto intensified through 2003, culminating with Prime Minister Chrétien's pledge to ratify the protocol that autumn. Finally my editors at UC Press gently, but firmly, insisted that if we were going to proceed with publication the ongoing revisions to the manuscript would have to end.

After the sound and fury of the battle before ratification, both sides refrained from issuing public comments in the early months of 2003, signalling that both industry and government were working on a compromise solution that would enable oil sands development to proceed without unduly harming either the industry's competitiveness or the government's commitment to environmental issues. Discussions between industry representatives and government officials culminated in a letter from the prime minister to the Canadian Association of Petroleum Producers in July 2003. In it, Chrétien basically pledged that the oil industry would remain competitive and outlined eight principles that would "guide the Government of Canada and the oil and gas sector in pursuing their climate change commitments." The principles included a pledge that no Canadian jobs would be lost as a result of a commitment to climate change initiatives, the efficient implementation of climate change policies, equitable treatment of all sectors of the Canadian economy, and special tax treatment for R&D.[1] Combined with the earlier pledge to limit the industry's Kyoto-related targets to no more than 15 per cent, this helped to ease investor's concerns.[2]

In the meantime, Kyoto received a series of body blows in May 2003 with word that ten of the fifteen states in the European Union had failed to meet emissions requirements under the protocol and that the EU's CO_2 emissions were actually on the rise.[3] Furthermore, in December 2003, Russian President Vladimir Putin's top economic adviser stated that Kyoto ran contrary to Russia's economic interests and suggested that his country should not ratify the treaty. Some estimates projected that President Putin's plan to double Russia's Gross Domestic Product by 2010 would push Russia's total CO_2 emissions to 104 percent of its 1990 levels, forcing Russia to become an emissions buyer instead of a seller.[4] The question of Russia's ratification was resolved in May 2004 when President Putin announced his country's willingness to ratify the protocol on the same day that the long-standing obstacles to Russian membership in the World Trade Organization were resolved, prompting some speculation that the two issues were linked.[5]

While environmental protection, energy conservation, and fuel efficiency are all parts of the equation, they are not the only factors to influence the direction of Canadian energy policy. Future Canadian governments will have to consider the fact that hampering domestic production could mean even higher energy prices for Canadians given the state of the world energy market. World crude prices nudged $50 per barrel in the summer of 2004 because of tight world supplies and growing demand.[6] While world energy prices have been volatile before, recent price spikes reflect a new development in world crude markets as North America now finds itself competing for world supplies against new global competitors, particularly from China and India. As respected industry analyst Daniel Yergin noted, even with enhanced conservation, increased consumption in Asia could mean that world consumption in ten years might exceed 115 million barrels per day compared to the current 81 million barrels per day.[7] Given questions about Saudi Arabia's ties to the September 11 plotters raised by the 9/11 Commission Report, guaranteeing North America a safe, secure, and reliable supply of energy is not a trivial matter.[8] The oil sands remain an important arrow in North America's energy quiver that could ultimately supplant the Middle East.

While no one – particularly a historian – can predict the future, studying the course of oil sands developments over the past century suggests that the range of issues and problems confronting the future of the industry are complex and multi-faceted. But one thing that immediately becomes apparent is how many of the same issues that confronted development in the early history of the oil sands affect the pace and scale of development to this day. Problems attracting a reliable labour force to Athabasca plagued Sidney Ells's attempts to collect samples from the deposit almost as surely as they do for current producers seeking to expand plant facilities, carry out maintenance, or ramp up production.[9] The same is true with the formulation and announcement of royalty and taxation regimes that will enable development to be profitable and beneficial for both the public and private sectors. Beyond these "nuts-and-bolts" issues, the size and scope of Alberta's oil sands industry remains inextricably linked to the United States and the world price of oil. Access to American markets and investment capital played an important role in the development of the early oil sands industry and may again prove decisive if the United States moves away from Middle Eastern oil.

BIBLIOGRAPHY

Archival Sources

Alberta Energy Utilities Board (Calgary, Alberta)
Hearings
Reports
Vertical File
Jimmy Carter Presidential Library (Atlanta, Georgia)
White House Central Files
Domestic Policy Papers
Office of the White House Counsel
Glenbow Archives (Calgary, Alberta)
Canadian Petroleum Association Fonds
General Files
Oil Sands and Heavy Oils Committee
Charles S. Lee Fonds
Helen and Mort Freeman Fonds
Independent Petroleum Association of Canada Fonds
Petroleum History Oral History Project
George Govier Interview
Provincial Archives of Alberta (Edmonton, Alberta)
Provincial Marketing Board
Premier's Papers
Ernest Manning
Karl A. Clark Papers
Department of Energy and Natural Resources
Papers of the Minister of Energy and Natural Resources
Department of Industry and Tourism
University of Alberta Archives (Edmonton, Alberta)
Sidney Martin Blair Papers

Government Documents

Advisory Committee on Heavy Oil and Oil Sands Development. *Annual Report, 1990*. Edmonton: Government Printer, 1990.

———. *Annual Report, 1989*. Edmonton: Government Printer, 1989.

———. *Annual Report, 1988*. Edmonton: Government Printer, 1988.

———. *Fourth Annual Report, 1987*. Edmonton: Government Printer, 1987.

———. *Third Annual Report, 1986*. Edmonton: Government Printer, 1986.

———. *Second Annual Report, 1985*. Edmonton: Government Printer, 1985.

———. *First Annual Report, 1984*. Edmonton: Government Printer, 1984.

Alberta Energy and Utilities Board. "Historical Overview of the Fort McMurray Area and Oil Sands Industry in Northeast Alberta." Earth Sciences Report #2000–05. Edmonton: Government Printer, 2000.

———. "Alberta's Energy Resources: 1996 in Review." Calgary: Government Printer, 1997.

Alberta Federal and Intergovernmental Affairs. "The Alberta Oil Sands Story." Edmonton: Government Printer, 1974.

Alberta Oil and Gas Conservation Board. Report 69-B, "Supplement to the Report of An Application of Atlantic Richfield Co., Cities Service Athabasca, Inc., Imperial Oil, and Royalite Company Limited under Part VIA of the Oil and Gas Conservation Act." Calgary: Government Printer, 1969.

———. Report 68-C, "Report of An Application of Atlantic Richfield Co., Cities Service Athabasca, Inc., Imperial Oil, and Royalite Company Limited Under Part VIA of the Oil and Gas Conservation Act." Calgary: Government Printer, 1968.

———. "Oil Sands Development Policy." Calgary: Government Printer, 1966.

———. Report 64-3. "Report to the Lieutenant Governor in Council With Respect to the Application of Great Canadian Oil Sands Limited Under Part VIA of the Oil and Gas Conservation Act." Edmonton: Government Printer, 1964.

———. Report 63-8. "Report to the Lieutenant Governor in Council With Respect to the Application of Cities Service and Shell Under Part VIA of the Oil and Gas Conservation Act." Edmonton: Government Printer, 1963.

———. Report 62-7. "Report to the Lieutenant Governor in Council With Respect to the Application of Great Canadian Oil Sands Limited Under Part VIA of the Oil and Gas Conservation Act." Edmonton: Government Printer, 1962.

———. Report 60-6. "Report to the Lieutenant Governor in Council With Respect to the Application of Great Canadian Oil Sands Limited Under Part VIA of the Oil and Gas Conservation Act." Edmonton: Government Printer, 1960.

Alberta Energy Resources and Conservation Board. "Forecast of the Supply and Requirements of Crude Oil, Synthetic Crude Oil and Pentanes Plus in Alberta, 1978–1995." Calgary: Government Printer, 1978.

Alberta Energy Utilities Board. "Historical Overview of the Fort McMurray Area and Oil Sands Industry in Northeast Alberta," Earth Sciences Report 2000-05. Edmonton: Government Printer, 2000.

———. "Alberta's Reserves 2000 and Supply/Demand Outlook 2001–2010." Calgary: Government Printer, 1998.

———. "Alberta's Energy Resources: 1997 in Review." Edmonton: Government Printer, 1998.

Alberta, Minister of the Environment. "Address to the Engineering Institute of Canada Conference: Development of the Alberta Oil Sands." Edmonton: April 17, 1974.

Alberta Oil Sands Environmental Research Program, *Interim Report to 1978*. Edmonton: Government Printer, 1979.

———. *First Annual Report, 1975*. Edmonton: Government Printer, 1976

Alberta Oil Sands Technology and Research Authority: *AOSTRA: A 15-Year Portfolio of Achievement*. Edmonton: Government Printer, 1991.

———. *Proceedings of 2nd Alberta Oil Sands University Seminar*. December 6, 7, 1976.

Alberta Research Council. *Thirty-Fourth Annual Report, 1953*. Edmonton: Government Printer, 1954.

———. *Thirty-Second Annual Report, 1951*. Edmonton: Government Printer, 1952.

———. *Thirty-First Annual Report, 1950*. Edmonton: Government Printer, 1951.

———. *Thirtieth Annual Report, 1949*. Edmonton: Government Printer, 1950.

———. *Twenty-Eighth Annual Report, 1947*. Edmonton: Government Printer, 1948.

———. *Annual Report, 1944*. Edmonton: Government Printer, 1945.

Alberta Technical Committee. "Report to the Minister of Mines and Minerals and the Oil and Gas Conservation Board." Edmonton: King's Printer, 1959.

Allen, F.H. "The Canadian Oil Sands: Race Against the Clock." *The United Nations Institute for Training and Research: The Future of Heavy Crude Oils and Tar Sands*. First International Conference, June 4–12, 1979. 29–32.

Blair, Sidney M. "Report on the Alberta Bituminous Sands." Edmonton: Government Printer, 1950.

Board of Trustees, Oil Sands Project. *Proceedings: Athabasca Oil Sands Conference*. Edmonton: King's Printer, 1951.

Breen, David H. Editor. *Selected Documents Pertaining to Natural Resources Ownership and Jurisdiction in Western Canada*. 4 Volumes. Vancouver: University of British Columbia Press, 1983.

Canada. *Royal Commission on Energy, Second Report*. Ottawa: Government Printer, 1959.

————. *Royal Commission on Energy, First Report*. Ottawa: Government Printer, 1958.

Canada Department of Mines and Resources. "Drilling and Sampling of Bituminous Sands of Northern Alberta." Volume I, *Results of Investigations, 1942–1947*. Ottawa: Government Printer, 1949.

————. "Preliminary Report on the Bituminous Sands of Northern Alberta." Ottawa: Government Printer, 1914.

Canada Energy Options Advisory Committee. *Energy and Canadians Into the 21st Century: A Report on the Energy Options Process*. Ottawa: Government Printer, 1988.

Canada Minister of Supply and Services. *The National Energy Program*. Ottawa: Government Printer, 1980.

Canada National Energy Board, *National Energy Board Annual Report, 1980*. Ottawa: Government Printer, 1981.

Carrigy, M.A. Editor. *The K.A. Clark Volume: A Collection of Papers on the Athabasca Oil Sands*. Alberta Research Council. Edmonton: Government Printer, 1963.

Clark, Karl A. "Athabasca Oil Sands: Historical Review and Summary of Technical Data." Alberta Research Council Contribution #69. Edmonton: Government Printer, 1957

————. "Athabasca Bituminous Sands." Alberta Research Council Contribution #24. Edmonton: Government Printer, 1951.

————. "Commercial Development of Alberta Bituminous Sands Now Economically Feasible." Alberta Research Council. Edmonton: Government Printer, 1951.

————. "Matters Needing Attention When the Bitumount Plant is Readied for Further Operation." Alberta Research Council. Edmonton: Government Printer, 1950.

————. "The Oil-Sand Separation Plant at Bitumount." Alberta Research Council Contribution #13. Edmonton: Government Printer, 1948.

————. "Hot-Water Separation of Alberta Bituminous Sand." Alberta Research Council Contribution #4. Edmonton: Government Printer, 1944.

————. "Introduction to Consolidated Mining and Smelting Company Limited Report on Blasting Oil Sands at Abasand Oils Ltd." Alberta Research Council. Edmonton: Government Printer, 1942.

————. "The Bituminous Sands of Alberta." Alberta Research Council Report #18. Edmonton: Government Printer, 1929.

————. "Bituminous Sand Development," *The Press Bulletin*, February 15, 1929.

Clark, Karl A. and D.S. Pasternack. "The Role of Very Fine Mineral Matter in the Hot Water Separation Process as Applied to the Athabasca Bituminous Sand." Alberta Research Council. Edmonton: Government Printer, 1949.

Government of Alberta. *Revised Statutes of Alberta*. Edmonton: Government Printer, various dates.

Ells, Sidney C. "Recollections of the Development of the Athabasca Oil Sands." Canada Department of Mines and Technical Surveys. Ottawa: Government Printer, 1962.

Frey John W. and H.C. Ide, eds. *A History of the Petroleum Administration For War: 1941–1945.* Washington: Government Printing Office, 1946.

Janisch, A. "Oil Sands and Heavy Oil: Can they Ease the Energy Shortage?" *The United Nations Institute for Training and Research: The Future of Heavy Crude Oils and Tar Sands.* First International Conference, June 4–12, 1979. 33–41.

Oil Sands Project, "Report to the Board of Trustees of the Oil Sands Project From Inception to December 31, 1948," Sessional Paper #53. Edmonton: King's Printer, 1949.

National Energy Board. "Canada's Oil Sands: A Supply and Market Outlook to 2015." Ottawa: Government Printer, 2000.

National Oil Sands Task Force on Oil Sands Strategies. "The Oil Sands: A New Energy Vision for Canada, Comprehensive Report." Edmonton: Government Printer, 1995.

———. "Canada's Oil Sands Industry: Yesterday, Today and Tomorrow." Edmonton: Government Printer, 1995.

———. "Securing a Sustainable Future for Canada's Oil Sands Industry." Edmonton: Government Printer, 1995.

Corporate Documents

Baugh, J.E. and L.L. McClennon. "The Future of Oilsands Mining Projects." *26th Canadian Chemical Engineering Conference.* October 3–6, 1976. 1–14.

Canadian Association of Petroleum Producers. "Climate Change: The Upstream Oil and Natural Gas Industry's Contribution to Canada's Debate on Climate Change and the Kyoto Protocol." Calgary: CAPP, 2002.

Canadian Petroleum Association. "An Assessment of Royalty Treatment and Other Factors Impacting Oil Sands Development." February, 1977.

———. "Submission of the Canadian Petroleum Association to the Government of Alberta Oil Sands Policy and Incentives." July 2, 1974.

———. "Report of Meeting Between Alberta Government Oil Sands Study Group and Industry," April 29, 1974.

———. "Submission to the National Energy Board in the Matter of Exportation of Oil." December, 1973.

———. "Submission to the Government of the Province of Alberta in the Matter of Royalty to be Applied to the Oil Sands." December 15, 1972.

———. "Submission to the Government of the Province of Alberta in the Matter of the Oil Sands Development Policy." November 9, 1971.

Canadian Petroleum Association and Independent Petroleum Association of Canada. "Joint Response by the Canadian Petroleum Association and the Independent Petroleum Association of Canada to May 24, 1974 Proposals of the Oil Sands Study Group."

———. "Joint Brief to the Oil Sands Study Group." May 6, 1974.

Canadian Resourcecon Limited. "Decision Making in the North: Oil Sands Case Study." Ottawa: Science Council of Canada, 1974.

Eglington, Peter and Maris Uffelmann. "An Economic Analysis of Oilsands Policy in Canada – The Case of Alsands and Wolf Lake," Economic Council of Canada Paper #259. Ottawa: Economic Council of Canada, 1984.

Esso Resources Canada Limited, "Final Environmental Impact Assessment," Vol. 1, *Project Description*, 1979.

Great Canadian Oil Sands, *Annual Report, 1974.*

Humphreys, Reginald D., F.K. Spragins, and D.R. Craig. "Oil Sands: Canada's First Answer to the Energy Shortage. *9th World Petroleum Congress.* May 11–16, 1975.

McKitrick, Ross, and Randal M. Wigle. "The Kyoto Protocol: Canada's Risky Rush to Judgment." Toronto: C.D. Howe Institute, 2002.

Perkins, Jody M. "Economic State of the US Oil and Natural Gas Exploration and Production Industry: Long-Term Trends and Recent Events." April 30, 1999. Washington: The American Petroleum Institute, 1999.

Walwyn, Stodgell & Company. "A Look At World Energy: The Athabasca Tar Sands." Toronto: Walwyn, Stodgily & Co., 1973.

Walter J. Levy & Associates "U.S. Import Policy: Implications for Canadian Export of Conventional and Tar Sand Oil," July 1968.

Websites

Alberta Energy Utilities Board. Available at *http://www.eub.gov.ab.ca/*

Alberta Research Council. Available at *http://www.arc.ab.ca/*

American Petroleum Institute. Available at *http://www.api.org*

Canadian Association of Petroleum Producers.
Available at *http://www.capp.ca*

C.D. Howe Institute. Available at *http://www.cdhowe.org*

David Suzuki Foundation. Available at *http://www.davidsuzuki.org*

Environment Canada. Available at *http://www.ec.gc.ca*

Greenpeace Canada. Available at *http://www.greenpeace.ca*

Government of Canada. Available at *http://www.gc.ca*

National Energy Board. Available at *http://www.neb.gc.ca/index_e.htm*

Prime Minister of Canada. Available at *http://pm.gc.ca/*

Syncrude Canada. Available at *http://www.syncrude.ca*

United States Department of Energy. Available at *http://www.doe.gov*

United States Energy Information Agency.
    Available at *http://www.eia.doe.gov*
The White House. Available at *http://www.whitehouse.gov*

Memoir/Autobiography/Papers

Adair, Al "Boomer" and Frank Dolphin. *Boomer: My Life with Peter, Don, and Ralph*. Edmonton: Polar Bear Publishing, 1994.

Ball, Max W. "Development of Athabasca Oil Sands," *Canadian Institute of Mining and Metallurgy*. XLIV (1941): 58–91.

Carter, Jimmy *Keeping Faith*. New York: Bantam Books, 1982.

Chrétien, Jean. *Straight From the Heart*. Toronto: McClelland and Stewart, 1986.

Diefenbaker, John G. *One Canada*. 3 Volumes. Toronto: Macmillan, 1975.

Ford, Gerald R. *A Time to Heal: The Autobiography of Gerald R. Ford*. New York: Harper and Row, 1979.

Nixon, Richard M. *RN: The Memoirs of Richard Nixon*. New York: Grosset & Dunlap, 1978.

Sheppard, Mary Clark. Editor. *Oil Sands Scientist: The Letters of Karl A. Clark, 1930–1949*. Edmonton: University of Alberta Press, 1989.

Trudeau, Pierre Elliott. *Memoirs*. Toronto: McClelland & Stewart, 1993.

Interviews

Mr. Don Axford, July 1998
Mr. Larry Morrison, March 2002
Mr. G. Vaughn Monroe, December 2001
Honorable Patricia Nelson, September 2001
Mr. Paul Precht, March 2002
Senator Nicholas Taylor, November 2001

Newspapers

Calgary Albertan
Calgary Herald
Calgary Sun
Canadian Mining Journal
Chemical Week
Christian Science Monitor
Denver Post
Edmonton Bulletin
Edmonton Journal

Energy Day

Enhanced Energy Recovery News

Financial Times (London)

Financial Post

Globe and Mail

The Guardian (U.K.)

The Hill Times (Ottawa)

Houston Chronicle

Journal of Commerce

Macleans

Montreal Gazette

Nickle's Daily Oil Bulletin

New York Times

Oilweek

Ottawa Citizen

Petroleum Finance Week

Platt's Oilgram News

Regina Leader-Post

The Wall Street Journal

Washington Post

The Windsor Star

Toronto Star

Secondary Sources

Abernathy, M. Glenn, Dilys M. Hill, and Phil Williams. *The Carter Years: The President and Policy Making*. New York: St. Martin's Press, 1984.

Al-Chalabi, Fadhil. *OPEC at the Crossroads*. Oxford: Pergamon Press, 1989.

Aim, Alvin L. and Robert J. Weiner, eds. *Oil Shock: Policy Response and Implementation*. Cambridge: Ballinger Publishing Company, 1984.

Andrew, Christopher. *For the President's Eyes Only: Secret Intelligence and the American Presidency from Washington to Bush*. New York: HarperCollins, 1995.

Bamberg, James. *British Petroleum and Global Oil, 1950–1975: The Challenge of Nationalism*. Cambridge: Cambridge University Press, 2000.

Beigie, Carle E. and Alfred O. Hero, Jr., eds. *Natural Resources in U.S.-Canadian Relations*. II Volumes. Boulder: Westview Press, 1980.

Bergsten, Fred, Thomas Horst, and Theodore Moran, *American Multinationals and American Interests*. Washington: Brookings, 1978.

Bianco, Anthony. *The Reichmanns: Family, Faith, Fortune, and the Empire of Olympia and York*. Toronto: Random House Canada, 1997.

Bidwell, Percy W. *Raw Materials: A Study of American Policy*. New York: Harper & Brothers, 1958.

Bird, Kai. *The Chairman: John J. McCloy The Making of the American Establishment.* New York: Simon & Schuster, 1992.

Bliss, Michael. *Right Honourable Men: The Descent of Canadian Politics from Macdonald to Mulroney.* Toronto: HarperPerennial, 1995.

———. *A Canadian Millionaire: The Life and Business Times of Sir Joseph Flavelle, Bart., 1858–1939.* Toronto: Macmillan, 1978.

Bothwell, Robert. *Canada and the United States: The Politics of Partnership.* Toronto: University of Toronto Press, 1992.

———, Ian Drummond, and John English. *Canada Since 1945.* Revised Edition. Toronto: University of Toronto Press, 1989.

——— and William Kilbourn. *C.D. Howe: A Biography.* Toronto: McClelland and Stewart, 1979.

Bradley, Robert L. *Oil, Gas & Government: The US. Experience.* 2 Volumes. London: Rowman & Littlefield Publishers, Inc., 1996.

Breen, David H. *Alberta's Petroleum Industry and the Conservation Board.* Edmonton: University of Alberta Press, 1993.

Brewster, Lawrence G. and Michael E. Brown. *The Public Agenda: Issues in American Politics.* Third Edition. New York: St. Martin's Press, 1993.

Brinkley, Alan and Davis Dyer, eds. *The Reader's Companion to the American Presidency.* Boston: Houghton Mifflin, 2000.

Brinkley, Alan. *The End of Reform: Liberalism in Recession and War.* New York: Vintage Books, 1995.

Bromley, Simon. *American Hegemony and World Oil: The Industry, the State System and the World Economy.* University Park: The Pennsylvania State University Press, 1991.

Brown, Anthony Cave. *Oil, God, and Gold: The Story of Aramco and the Saudi Kings.* New York: Houghton Mifflin Company, 1999.

Bull-Berg, Hans Jacob. *American International Oil Policy: Causal Factors and Effect.* New York: St. Martin's Press, 1987.

Byrne, T.C. *Alberta's Revolutionary Leaders.* Calgary: Detselig Enterprises, 1991.

Campagna, Anthony S. *Economic Policy in the Carter Administration.* Westport: Greenwood Press, 1995.

Caplan, Gerald, Michael Kirby, and Hugh Segal. *Election: The Issues, the Strategies, the Aftermath.* Scarborough: Prentice-Hall, 1989.

Carroll, Peter N. *It Seemed Like Nothing Happened: The Tragedy and Promise of America in the 1970s.* New York: Holt, Rinehart, and Winston, 1982.

Chace, James. *Acheson: The Secretary of State Who Created the American World.* New York: Simon & Schuster, 1998.

Chernow, Ron. *Titan: The Life of John D. Rockefeller.* New York: Random House, 1998.

Chong, Ken P. and John Ward Smith, eds. *Mechanics of Oil Shale.* London: Elsevier Applied Science Publishers, 1984.

Clark, John G. *Energy and the Federal Government: Fossil Fuel Policies, 1900–1946.* Chicago: University of Illinois Press, 1987.

Clarkson, Stephen. *Uncle Sam and Us: Globalization, Neoconservatism, and the Canadian State.* Toronto: University of Toronto Press, 2002.

Clarkson, Stephen and Christina McCall. *Trudeau and Our Times: The Magnificent Obsession.* Volume I of 2 volumes. Toronto: McLelland & Stewart, 1990.

———. *Trudeau and Our Times: The Heroic Delusion.* Volume 2 of 2 volumes. Toronto: McLelland & Stewart, 1994.

Comfort, Darlene J. *The Abasand Fiasco: The Rise and Fall of a Brave Pioneer Oil Sands Extraction Plant.* Edmonton: Friesen Printers, 1980.

Cowhey, Peter F. *The Problems of Plenty: Energy Policy and International Politics.* Berkeley: University of California Press, 1985.

Creighton, Donald. *Canada's First Century.* Toronto: Macmillan, 1970.

Dabbs, Frank. *Ralph Klein: A Maverick Life.* Vancouver: Greystone Books, 1995.

Dallek, Robert. *FDR and United States Foreign Policy.* New York: Oxford University Press, 1995.

Dawson, R.M. *William Lyon Mackenzie King: A Political Biography, 1874–1923.* Toronto, University of Toronto Press, 1958.

Desse, David A. and Joseph S. Nye, eds. *Energy and Security: A Report of Harvard's Energy and Security Research Project.* Cambridge: Ballinger Publishing Company, 1981.

Doern, G. Bruce and Glen Toner, *The Politics of Energy: The Development and Implementation for the NEP.* Toronto: Methuen, 1984.

Doran, Charles F. *Forgotten Partnership: US-Canada Relations Today.* Baltimore: Johns Hopkins University Press, 1984.

Duffy, John. *Fights of Our Lives: Elections, Leadership, and the Making of Canada.* Toronto: Harper Collins, 2002.

Dumbrell, John. *The Carter Presidency: A Re-Evaluation.* Manchester: Manchester University Press, 1993.

Dumenil, Lynn. *Modern Temper: American Culture and Society in the 1920's.* New York: Hill and Wang, 1995.

Ebinger, Charles K. *The Critical Link: Energy and National Security in the 1980s.* Cambridge: Ballinger Publishing Company, 1982.

Eckes, Alfred E., Jr. *Opening America's Market: US Foreign Trade Policy Since 1776.* Chapel Hill: University of North Carolina Press, 1995.

———. *The United States and the Global Struggle for Minerals.* Austin: University of Texas Press, 1979.

———. *A Search for Solvency: Bretton Woods and the International Monetary System, 1941–1971.* Austin: University of Texas Press,1975.

English, John. *The Life of Lester Pearson: The Worldly Years.* Toronto: Alfred A. Knopf, 1992.

———. *The Decline of Politics: The Conservatives and the Party System*. Toronto: University of Toronto Press, 1977.

Erickson, Edward W. and Leonard Waverman, eds. *The Energy Question: An International Failure of Policy*. 2 Volumes. Toronto: University of Toronto Press, 1974.

Feis, Herbert. *Petroleum and American Foreign Policy*. Stanford: Food Research Institute, 1944.

Ferguson, Barry Glen. *Athabasca Oil Sands: Northern Resource Exploration, 1875–1951*. Edmonton: Alberta Culture, 1985.

Fink, Gary M. and Hugh Davis Graham. *The Carter Presidency: Policy Choices in the Post-New Deal Era*. Lawrence: University Press of Kansas, 1998.

Finkel, Alvin. *The Social Credit Phenomenon in Alberta*. Toronto: University of Toronto Press, 1989.

Fitzgerald, J. Joseph. *Black Gold With Grit*. Sidney: Gray's Publishing, 1978.

Fossum, John Erik. *Oil, the State, and Federalism: The Rise and Demise of PetroCanada as a Statist Impulse*. Toronto: University of Toronto Press, 1997.

Foster, Peter. *Self Serve: How Petro-Canada Pumped Canadians Dry*. Toronto: Macfarlane, Walter, & Ross, 1992.

———. *Other People's Money: The Banks, the Government and Dome*. Toronto: Collins, 1983.

———. *Sorcerer's Apprentices: Canada's Super-Bureaucrats and the Energy Mess*. Toronto: HarperCollins, 1982.

———. *Blue-Eyed Sheiks: The Canadian Oil Establishment*. Toronto: Totem Books, 1979.

Fotheringham, Allan. *Malice In Blunderland*. Toronto: Bantam Books, 1982.

Freeman, J.M. *Biggest Sellout in History: Foreign Ownership of Alberta's Oil and Gas Industry and the Oil Sands*. Edmonton: Alberta New Democratic Party, 1966.

Friedland, Edward, Paul Seabury and Aaron Wildavsky. *The Great Détente Disaster: Oil and the Decline of American Foreign Policy*. New York: Basic Books, 1975.

Friedman, Thomas L. *The Lexus and the Olive Tree: Understanding Globalization*. New York: Anchor Books, 2000.

Gaddis, John L. *We Now Know: Rethinking Cold War History*. New York: Oxford University Press, 1997.

Ghosh, Arabinda. *OPEC, The Petroleum Industry, and United States Energy Policy*. London: Quorum Books, 1983.

Gibb, George S. and Evelyn H. Knowlton. *A History of Standard Oil: The Resurgent Years, 1911–1927*. New York: Harper, 1956.

Gibbins, Roger. *Conflict and Unity: An Introduction to Canadian Political Life*. Toronto: Methuen, 1985.

Giebelhaus, August W. *Business and Government in the Oil Industry: A Case Study of Sun Oil, 1876–1945*. Greenwich: Jai Press Inc., 1980.

Gillespie, Kate and Clement M. Henry, eds. *Oil in the New World Order*. Gainesville: University Press of Florida, 1995.

Gilpin, Robert *U.S. Power and the Multinational Corporation: The Political Economy of Foreign Direct Investmen*. New York: Basic, 1975.

Goodwin, Craufurd D. Editor. *Energy Policy in Perspective: Today's Problems, Yesterday's Solutions*. Washington: The Brookings Institution, 1981.

Graham, Ron *One-Eyed Kings: Promise and Illusion in Canadian Politics*. Toronto: Totem Books, 1986.

Granatstein, J.L. and Norman Hillmer. *For Better or For Worse: Canada and the United States to the 1990's*. Toronto: Copp Clark Pittman, 1991.

Granatstein, J.L. and Robert Bothwell. *Pirouette: Pierre Trudeau and Canadian Foreign Policy*. Toronto: University of Toronto Press, 1990.

Granatstein, J.L., Irving M. Abella, David J. Bercuson, R. Craig Brown, and H. Blair Neatby. *Twentieth Century Canada*, 2nd edition. Toronto: McGraw-Hill, 1986.

Greider, William. *One World; Ready or Not: The Manic Logic of Global Capitalism*. New York: Simon & Schuster, 1997.

Grose, Peter. *Gentleman Spy: The Life of Allen Dulles*. New York: Houghton Mifflin, 1994.

Guertin, Donald L., W. Kenneth Davis, and John E. Gray, eds. *US. Energy Imperatives for the 1990's: Leadership, Efficiency, Environmental Responsibility, and Sustained Growth*. New York: University Press of America, 1991.

Haas, Garland, A. *Jimmy Carter and the Politics of Frustration*. Jefferson, North Carolina: McFarland & Company, Inc., 1992.

Hamby, Alonzo. *Man of the People*. New York: Oxford University Press, 1995.

———. *Liberalism and Its Challengers: From F.D.R. to Bush*. Second Edition. New York: Oxford University Press, 1992.

Hanson, Eric J. *The Dynamic Decade: The Evolution and Effects of the Oil Industry in Alberta*. Toronto: McClelland and Stewart, 1958.

Hargrove, Erwin C. *Jimmy Carter as President: Leadership and the Politics of the Public Good*. Baton Rouge: Louisiana State University Press, 1988.

Hilborn, James D. ed. *Dusters and Gushers: The Canadian Oil and Gas Industry*. Toronto: Pitt Publishing Company Limited, 1968.

Hillmer, Norman and J.L. Granatstein. *From Empire to Umpire: Canada and the World to the 1990's*. Toronto: Copp Clark Longman, 1994.

Hoff, Joan. *Nixon Reconsidered*. New York: Basic Books, 1994.

Hogan, Michael J. *A Cross of Iron: Harry S. Truman and the Origins of the National Security State, 1945–1954*. New York: Cambridge University Press, 1998.

Horwich, George and David Leo Weimer, eds. *Responding to International Oil Crises*. Washington: American Enterprise Institute for Public Policy Research, 1988.

Hoskins, Halford L. *Middle East Oil in United States Foreign Policy*. Westport: Hyperion Press, 1950.

Hustak, Allan. *Peter Lougheed*. Toronto: McClelland and Stewart, 1979.

Ikenberry, G. John. *Reasons of State: Oil Politics and the Capacities of American Government*. Ithaca: Cornell University Press, 1988.

Isaacson, Walter. *Kissinger: A Biography*. New York: Simon & Schuster, 1992.

———. and Evan Thomas. *The Wise Men: Six Friends and the World They Made*. New York: Simon & Schuster, 1986.

Jeffrey, Brooke. *Hard Right Turn: The New Face of Neo-Conservatism in Canada*. Toronto: HarperCollins, 1999.

Johns, Walter H. *A History of the University of Alberta, 1908–1969*. Edmonton: University of Alberta Press, 1981.

Johnson, Arthur M. *The Challenge of Change: The Sun Oil Company, 1945–1977*. Columbus: Ohio State University Press, 1983.

Johnson, Haynes. *In the Absence of Power: Governing America*. New York: The Viking Press, 1980.

Jones, Charles O. *The Trusteeship Presidency: Jimmy Carter and the United States Congress*. Baton Rouge: Louisiana State University Press, 1988.

Karlsson, Svante. *Oil and the World Order: American Foreign Oil Policy*. Totowa: Barnes & Noble Books, 1986.

Kaufman, Burton I. *The Presidency of James Earl Carter, Jr*. Lawrence: University Press of Kansas, 1993.

———. *The Oil Cartel Case: A Documentary Study of Antitrust Activity in the Cold War Era*. Westport: Greenwood Press, 1978.

Kennedy, John de Navarre. *History of the Department of Munitions and Supply: Canada in the Second World War*. 2 volumes. Ottawa: King's Printer, 1950.

Kennedy, Paul. *Preparing for the Twenty-First Century*. New York: Vintage Books, 1994.

Kennedy, Tom. *Quest: Canada's Search for Arctic Oil*. Edmonton: Reidmore Books, 1988.

Kent, Marian. *Oil and Empire: British Policy and Mesopotamian Oil, 1900–1920* London: Macmillan, 1976.

Klapp, Merrie Gilbert. *The Sovereign Entrepreneur: Oil Policies in Advanced and Less Developed Capitalist Countries*. Ithaca: Cornell University Press, 1987.

Klare, Michael T. *Resource Wars: The New Landscape of Global Conflict*. New York: Henry Holt & Company, 2002.

Klassen, Henry C. *A Business History of Alberta* Calgary: University of Calgary Press, 1999.

Kohl, Wilfred L., ed. *After the Price Collapse: OPEC, The United States, and the World Oil Market*. Baltimore: The Johns Hopkins University Press, 1991.

Kraples, Edward N. *Oil Crisis Management: Strategic Stockpiling for International Security*. Baltimore: The Johns Hopkins University Press, 1980.

Krasner, Stephen D. *Defending the National Interest: Raw Materials Investments and U.S. Foreign Policy.* Princeton: Princeton University Press, 1978.

Kunz, Diane B. *Butter and Guns: America's Cold War Economic Diplomacy.* New York: The Free Press, 1997.

Landes, David S. *The Wealth and Poverty of Nations: Why Some are so Rich and Some so Poor.* New York: W.W. Norton & Company, 1998.

Leffler, Melvyn P. *A Preponderance of Power: National Security, the Truman Administration and the Cold War.* Stanford: Stanford University Press, 1992.

Leffler, Melvyn P. and David S. Painter, eds. *Origins of the Cold War: An International History.* New York: Routledge, 1994.

Levant, Ezra. *Fight Kyoto: The Plan To Protect Our Economic Future.* Calgary: JMCK Publishing, 2002.

Lisac, Mark. *The Klein Revolution.* Edmonton: NeWest Publishers, 1995.

Lovett, William A., Alfred E. Eckes, Jr., and Richard Brinkman, *U.S. Trade Policy: History, Theory, and the WTO.* New York: M.E. Sharpe, 1999.

MacDonald, L. Ian, ed. *Free Trade: Risks and Rewards.* Montreal: McGill-Queens University Press, 2000.

MacGregor, James G. *A History of Alberta.* Edmonton: Hurtig Publishers, 1981.

Mansell, Robert L. and Michael B. Percy. *Strength in Adversity: A Study of the Alberta Economy.* Edmonton: The University of Alberta Press, 1990.

Martin, Don. *King Ralph: The Political Life and Success of Ralph Klein.* Toronto: Key Porter Books, 2002.

Martin, Lawrence. *Chrétien: The Will to Win.* Toronto: Lester Publishing, 1995.

McCall-Newman, Christine. *Grits: An Intimate Portrait of the Liberal Party.* Toronto: Macmillan, 1982.

McKee, David L., ed. *Canadian-American Economic Relations: Conflict and Cooperation on a Continental Scale.* New York: Praeger Publishers, 1988.

McKenzie-Brown, Peter, Gordon Jaremko, and David Finch. *The Great Oil Age: The Petroleum Industry in Canada.* Calgary: Detselig Publishers, 1993.

Mikesell, Raymond F. and Hollis B. Chenery. *Arabian Oil: America's Stake in the Middle East.* Chapel Hill: University of North Carolina Press, 1949.

Milkis, Sidney M. *The President and the Parties: The Transformation of the American Party System Since the New Deal.* New York: Oxford University Press, 1949.

Miller, Aaron David. *Search for Security: Saudi Arabian Oil and American Foreign Policy, 1939–1949.* Chapel Hill: University of North Carolina Press, 1980.

Milward, Alan S. *War, Economy and Society, 1939–1945.* Berkeley: University of California Press, 1977.

Morton, W.L. *The Progressive Party in Canada.* Toronto: University of Toronto Press, 1950.

Nash, Gerald. *United States Oil Policy, 1890–1964: Business and Government in Twentieth Century America*. Pittsburgh: University of Pittsburgh Press, 1968.

Newman, Peter C. *The Canadian Revolution: From Deference to Defiance*. Toronto: Penguin Books, 1995.

Nikiforuk, Andrew, Sheila Pratt, and Don Wanagas, eds. *Running on Empty: Alberta After the Boom*. Edmonton: NeWest, 1987.

Nordhauser, Norman E. *The Quest for Stability: Domestic Oil Regulation, 1917–1935*. New York: Garland Publishing, 1979.

Norrie, Kenneth and Douglas Owram, *A History of the Canadian Economy*. Toronto: Harcourt Brace & Co., 1996.

Olien, Roger M. and Diana Davids Olien. *Oil and Ideology: The Cultural Creation of the American Petroleum Industry*. Chapel Hill: The University of North Carolina Press, 2000.

Ollinger, Michael. *Organizational Form and Business Strategy in the U.S. Petroleum Industry*. Maryland: University Press of America, 1993.

Overy, Richard. *Why the Allies Won*. London: Pimlico Press, 1995.

Pach, Chester J. Jr. and Elmo Richardson. *The Presidency of Dwight D. Eisenhower*. Revised Edition. Lawrence: University of Kansas Press, 1991.

Painter, David S. *Oil and the American Century: The Political Economy of US Foreign Oil Policy, 1941–1954*. Baltimore: The Johns Hopkins University Press, 1986.

Patterson, Thomas G. and Dennis Merrill, eds. *Major Problems in American Foreign Relations. Volume 2: Since 1914*. Fourth Edition. Lexington: D.C. Heath & Co., 1994.

Peters, B. Guy. *American Public Policy: Promise and Performance*. Third Edition. Chatham: Chatham House Publishers, 1993.

Philip, George. *The Political Economy of International Oil*. Edinburgh: Edinburgh University Press, 1994.

Pratt, Larry. *The Tar Sands: Syncrude and the Politics of Oil*. Edmonton: Hurtig Publishers, 1976.

Randall, Stephen J. *United States Foreign Oil Policy, 1919–1948: For Profits and Security*. Montreal: McGill-Queen's University Press, 1985.

Randall, Stephen J. and Herman W. Konrad, eds. *NAFTA in Transition*. Calgary: University of Calgary Press, 1995.

Rhodes, Richard. *The Making of the Atomic Bomb*. New York: Simon & Schuster, 1986.

Richards, John and Larry Pratt, *Prairie Capitalism: Power and Influence in the New West*. Toronto: McClelland and Stewart, 1979.

Robert, Maryse. *Negotiating NAFTA: Explaining the Outcome in Culture, Textiles, Autos, and Pharmaceuticals*. Toronto: University of Toronto Press, 2000.

Robinson, J.M.A. and J.Y. Jessup, *The Alsands Project*. Calgary: Canadian Major Projects Association, 1988.

Rosenbaum, Herbert D. and Alexej Ugrinsky, eds. *The Presidency and Domestic Policies of Jimmy Carter*. Westport: Greenwood Press, 1994.

Rosenbaum, Walter A. *Energy, Politics, and Public Policy*. Second Edition. Washington: Congressional Quarterly Press, 1987.

Rosenberg, Emily S. *Spreading the American Dream: American Economic and Cultural Expansion, 1890–1945*. New York: Hill and Wang, 1982.

Russell, Paul. L. *Oil Shales of the World: Their Origin, Occurrence & Exploitation*. Oxford: Pergamon Press, 1990.

Scheppach, Raymond C. and Everett M. Ehrlich, eds. *Energy-Policy Analysis and Congressional Action*. Lexington: D.C. Heath and Company, 1982.

Schlesinger, James R. *American Security and Energy Policy*. Manhattan: Kansas State University, 1980.

Shwadran, Benjamin. *The Middle East, Oil and the Great Powers*. New York: John Wiley & Sons, 1973.

Smith, Philip. *The Treasure-Seekers: The Men Who Built Home Oil*. Toronto: Macmillan Canada, 1979.

Stobaugh, Robert and Daniel Yergin, eds. *Energy Future: Report of the Energy Project at the Harvard Business School*. New York: Random House, 1979.

Stoff, Michael B. *Oil, War, and American Security: The Search for a National Policy on Foreign Oil, 1941–1947*. New Haven: Yale University Press, 1980.

Szyliowicz, Joseph S. and Bard E. O'Neill, eds. *The Energy Crisis and U.S. Foreign Policy*. New York: Praeger Publishers, 1975.

Thompson, John Herd and Stephen J. Randall. *Canada and the United States: Ambivalent Allies*. Athens: University of Georgia Press, 1994.

Thompson, Kenneth W., ed. *The Carter Presidency: Fourteen Intimate Perspectives of Jimmy Carter*. New York: University Press of America, 1982.

Tempest, Paul, ed. *The Politics of Middle East Oil*. London: Granham & Trotman, 1993.

Trager, Frank N., ed. *Oil, Divestiture and National Security*. New York: Crane Russak & Company, Inc., 1977.

Usalner, Eric M. *Shale Barrel Politics: Energy and Legislative Leadership*. Stanford: Stanford University Press, 1989.

Verleger, Philip K., Jr. *Adjusting to Volatile Energy Prices*. Washington: Institute for International Economics, 1993.

Vernon, Raymond *Sovereignty at Bay: The Spread of U.S. Enterprises*. New York: Basic, 1971.

Vietor, Richard K. *Energy Policy in America since 1945: A Study of Business-Government Relations*. Cambridge: Cambridge University Press, 1984.

Watkins, G. Campbell, ed. *PetroMarkets: Probing the Economics of Continental Energy*. Vancouver: The Fraser Institute, 1989.

Watkins, G. Campbell and M.A. Walker, eds. *Reaction: The National Energy Program*. Vancouver: The Fraser Institute, 1981.

Williamson, Harold F. *The American Petroleum Industry: The Age of Energy, 1899–1959*. (Evanston: Northwestern University Press, 1963.

Wood, David G. *The Lougheed Legacy*. Toronto: Key Porter Books, 1985.

Yen, Tehfu, ed. *Oil Shale*. London: Elsevier Scientific Publishing Company, 1976.

Yergin, Daniel. *The Prize: The Epic Quest for Oil, Money and Power*. New York: Simon & Schuster, 1992.

Yergin, Daniel and Joseph Stanislaw. *The Commanding Heights: The Battle Between Government and the Marketplace that is Remaking the Modern World*. New York: Simon & Schuster, 1998.

Journal Articles

Ahrari, Mohmmed E. "A Paradigm of 'Crisis' Decision Making: The Case of Synfuels Policy," *British Journal of Political Science* 17 (1985): 71–91.

Akins, James E. "The Oil Crisis: This Time the Wolf is Here." *Foreign Affairs* 51 (1973): 462–90.

Amuzegan, Jahangir. "The Oil Story: Facts, Fiction and Fair Play." *Foreign Affairs* 51 (1973): 676–89.

Beaver, William. "The U.S. Failure to Develop Synthetic Fuels in the 1920's," *The Historian* 53 (1991): 241–54.

Binnema, Theodore. "Making Way for Canadian Oil: United States Policy, 1947–59," *Alberta History* 45 (1997): 15–23.

Brandie, G.W., R.H. Clark and J.C. Wiginton. "The Economic Enigma of the Tar Sands," *Canadian Public Policy* VIII (1982): 156–64.

Breen, David H. "Anglo-American Rivalry and the Evolution of Canadian Petroleum Policy to 1930," *Canadian Historical Review* LXII (1981): 283–303.

Campbell, John C. "Oil Power in the Middle East." *Foreign Affairs* 56 (1978): 89–110.

Crow, Michael. M. and Gregory L. Hager. "Political Versus Technical Risk Deduction and the Failure of U.S. Synthetic Fuel Development Efforts." *Policy Studies Review* 5 (1986): 145–52.

Cuff, Robert and J.L. Granatstein, "The Rise and Fall of Canadian-American Free Trade, 1947–1948," *Canadian Historical Review* 57 (4)

Curtis, John M. and Andrew R. Moroz. "Canada-United States Trade and Policy Issues: Towards a New Relationship?" *Canadian Public Policy* VIII (1982): 405–7.

Daniel, Terrence E. and Henry M. Goldberg. "Moving Towards World Pricing for Oil and Gas," *Canadian Public Policy*. VIII (1982): 3–13.

Empey, W.F. "The Impact of Higher Energy Prices in Canada." *Canadian Public Policy*. VII (1981): 28–34.

Enemark, Tex. "A Federal-Provincial Affairs Perspective." *Canadian Public Policy*, VIII (1982): 40–44.

Feick, John. E. "Prospects for the Development of Minable Oil Sands." *Canadian Public Policy* IX (1983): 297–303.

Feis, Herbert. "Oil for Peace or War," *Foreign Affairs* 32 (1944): 416–39.

Govier, George. "Alberta's Synthetic Crude Oil Development Policy," *Quarterly of the Colorado School of Mines* 65 (1970): 229–41.

Grey, Rodney de C. "Some Issues in Canada-US Trade Relations." *Canadian Public Policy* VIII (1982): 451–59.

Healey, Denis. "Oil, Money & Recession," *Foreign Affairs* 58 (1980): 217–30.

Helliwell, John F. "The National Energy Conflict." *Canadian Public Policy* VII (1981): 15–23

———. "Taxation of Oil and Gas Revenues of Four Countries: Canada." *The Energy Journal* 3 (1983): 20–31.

Helliwell, John F. and Robert N. McRae. "Resolving the Energy Conflict: From the National Energy Program to the Energy Agreements." *Canadian Public Policy* VIII (1982): 14–23.

Hume. O.S. "Results and Significance of Drilling Operations in the Athabaska Bituminous Sands." *Canadian Institute of Mining and Metallurgy*. L (1950): 298–333.

Jenkins, Barbara. "Reexamining the 'Obsolescing Bargain': A Study of Canada's National Energy Program," *International Organization* 40 (1986): 139–65.

Lambright, W. Henry, Michael Crow and Ralph Shangraw. "National Projects in Civilian Technology." *Policy Studies Review* 3 (1986): 73–92.

Levy, Walter J. "World Oil: Cooperation or International Crisis." *Foreign Affairs* 52 (1974): 690–713.

Maclaren, Roy. "Canadian Views on the US Government Reaction to the National Energy Program." *Canadian Public Policy* VIII (1982): 493–97.

McRae, Robert N. "A Major Shift in Canada's Energy Policy: Impact of the National Energy Program." *The Journal of Energy and Development* (Spring, 1982): 173–98.

Page, Donald. "An Energy Crisis in Reverse: Canada As Net Oil Importer." *International Perspectives*. 2 (1974): 18–21.

Pearse, Chas. R. "Athabasca Tar Sands." *Canadian Geographical Journal* 76 (1968): 2–9.

Pollack, Gerald A. "The Economic Consequences of the Energy Crisis." *Foreign Affairs* 52 (1974): 452–71.

Scarfe, Brian L. "Canadian Energy Prospects: Natural Gas, Tar Sands and Oil Policy," *Contemporary Policy Issues* III (1985): 13–24.

———. "The Federal Budget and Energy Program, October 28th, 1980: A Review." *Canadian Public Policy* VII (1981): 1–14

Stivers, William. "International Politics and Iraqi Oil, 1918–1928: A Study in Anglo-American Diplomacy." *Business History Review* 55 (1981): 517–40.

Taylor, Graham D. "Sun Oil Company and the Great Canadian Oil Sands Ltd.: The Financing and Management of a 'Pioneer' Enterprise, 1962–1974," *Journal of Canadian Studies* 20 (1985): 102–21.

Uffelmann, Maris. "Hydrocarbon Supply Costs," *Canadian Public Policy* XI (1985): 397–401.

Vietor, Richard H.K. "The Synthetic Liquid Fuels Program: Energy Politics in the Truman Era," *Business History Review* LIV (1980): 1–34.

Wahby, Mandy J. "Petroleum Taxation and Efficiency: The Canadian System in Question." *Journal of Energy and Development.* 9 (1984): 111–27.

Wonder, Edward. "The US Government Response to the Canadian National Energy Program." *Canadian Public Policy* VIII (1982): 480–93.

Theses/Dissertations

Dooley, Alan Michael. "The Cold Lake Oil Sands Project: An Economic Analysis." Master's Thesis, University of Alberta, 1981.

Gloin, Kevin James. "Canada's Role in the United States National Security Policy: Strategic Materials in the Eisenhower-Diefenbaker Era." Master's Thesis, University of Calgary, 1995.

Goerzen, Brenda. "Free Trade and Petrochemicals in Alberta." Unpublished M.B.A. Thesis, University of Calgary, 1988.

Greene, William Noel. "Strategies of the Major Oil Companies." Unpublished Ph.D. Dissertation, Harvard University, 1982.

Kohler, Larry R. "Canadian-American Oil Diplomacy: The Adjustment of Conflicting National Oil Policies, 1955–1973." Unpublished Ph.D. Dissertation, The Johns Hopkins University, 1983.

Sievwright, Eric Colville. "The Effect of Petroleum Development on the Alberta Economy, 1947–1957." Unpublished Ph.D. Dissertation, McGill University, 1960.

NOTES

Introduction

1   Energy Information Agency. "Canada Energy Oil Information." February 2001. [Online]: <http://www.eia.doe.gov/emeu/cabs/canada.html>[3 August 2001].

2   Daniel Yergin, "Gulf Oil – How Important is it Anyway?" *The Financial Times* (London), March 22, 2003, p. 1.

3   In its raw state, bitumen is an asphalt-like oil that requires upgrading to become transportable by pipeline and usable in conventional refineries. The terms "oil sands," "tar sands" and "bituminous sands" have all been used at various times to describe the same material and are used interchangeably throughout the text.

4   Richard W. Phelps, "Producing Oil the Old-Fashioned Way; Mining It," *Engineering and Mining Journal*, December 1997, p. 18; James Brooke, "Canada Is Unlocking Petroleum From Sand," *The New York Times*, January 23, 2001, [Online]: <http://www.nythimes.com/2001/01/23/business/23SAND.html> [2 August 2001].

5   Clyde H. Farnsworth, "Unlocking Oil in Canada's Tar Sands," *The New York Times*, December 28, 1994. [Online]: <http://web.lexis-nexis.com/universe/documentA1&-md=d8d17e0907792046f53afa2b54a3b03a7> [2 September 2000].

6   Hugh G. J. Aitken, "Defensive Expansionism: The State and Economic Growth in Canada," *The State and Economic Growth*, p. 113.

7   John Erik Fossum, *Oil, the State, and Federalism* (Toronto: University of Toronto Press, 1996), pp. 37–38.

8   Diane B. Kunz, *Butter and Guns: America's Cold War Economic Diplomacy* (New York: Free Press, 1997), p. 225.

9   Farnsworth, "Unlocking Oil in Canada's Tar Sands."

10 Ronald A. Shearer, "Nationality, Size of Firm, and Exploration for Petroleum in Western Canada, 1946–1954," *The Canadian Journal of Economics and Political Science*, 30 (1964), pp. 211–27; Alexander Dow, "Finance and Foreign Control in Canadian Base Metal Mining,

1918–1955," *Journal of Economic History* 44 (1984), p. 56; Douglas E. Cass, "Investment in Alberta Petroleum, 1912–1930," Unpublished Masters Thesis, University of Calgary, 1985.

11   Energy Information Agency/International Energy Agency, *International Energy Outlook, 2002* (Washington, 2002), p. 11. Available [Online]: < http://www.eia.doe.gov/oiaf/ieo/pdf/consumption.pdf>   [14   April 2003].

12   Ibid., p. 105. Available [Online]: <http://www.eia.doe.gov/oiaf/ieo/pdf/ hydroelectricity.pdf> [14 April 2003].

13   U.S. Department of Energy, *Fuel Cell Report to Congress (ESECS EE-1973)* (Washington: Government Printer, 2003) Available [Online]: <http://www.eere.energy.gov/hydrogenandfuelcells/pdfs/fc_report_congress_feb2003.pdf > [14 April 2003].

Chapter 1

1   David H. Breen, *Alberta's Petroleum Industry and the Conservation Board* (Edmonton: University of Alberta Press, 1993), p. 8.

2   William Beaver, "The Failure to Develop Synthetic Fuels in the 1920s," *Historian* 53 (2), pp. 241–54.

3   The situation would not be resolved until after the Second World War when possession of the oil sands region quietly reverted to the province with little fanfare.

4   Oil Sands Discovery Centre, "History Timeline, 1700–1919." [Online]: <http://collections.ic.gc.ca/oil/index1.htm> [13 August 2002].

5   Breen, *Alberta's Petroleum Industry*, p. 8.

6   Mary Clark Sheppard, ed., *Oil Sands Scientist: The Letters of Carl A. Clark, 1930–1949* (Edmonton: University of Alberta Press, 1989), p. 13.

7   Jack Granatstein et al., *Twentieth Century Canada*, 2d ed. (Toronto: McGraw-Hill, 1986), p. 42.

8   RG33/52 in David H. Breen ed., *Selected Documents Pertaining to Natural Resources Ownership and Jurisdiction in Western Canada* (Vancouver: University of British Columbia Press, 1983), vol. 3 of 4 vols.

9   Marian Kent, *Oil and Empire: British Policy and Mesopotamian Oil, 1900–1920* (London: Macmillan, 1976), pp. 36; David H. Breen, "Canadian Petroleum Policy," *Canadian Historical Review* LXII (1981), p. 285.

10   Philip Smith, *The Treasure-Seekers: The Men Who Built Home Oil* (Toronto: Macmillan, 1979), p. 19.

11   Ibid.

12   Darlene J. Comfort, *The Abasand Fiasco: The Rise and Fall of a Brave Pioneer Oil Sands Extraction Plant* (Edmonton: Friesen Printers, 1980), p. 46.

13   Henry C. Klassen, *A Business History of Alberta* (Calgary: University of Calgary Press, 1999), p. 73.

14   Ibid., p. 78.

15   Walter H. Johns, *A History of the University of Alberta, 1908–1969* (Edmonton: University of Alberta Press, 1981).

16   Sheppard, *Oil Sands Scientist*, p. 15.

17   Clayton Jones, "Alberta. Canada's Energy Giant, Turns Sand into Black Gold," *The Christian Science Monitor*, November 20, 1980, p. 3.

18   Breen, "Canadian Petroleum Policy," p. 285.

19   Ibid., 286.

20   Breen, *Alberta's Petroleum Industry*, p. 288.

21   Sidney C. Ells, "Recollections of the Development of the Athabasca Oil Sands," (Ottawa: Government Printer, 1966), p. 3.

22   Comfort, *The Abasand Fiasco*, pp. 20–23.

23   Sheppard, *Oil Sands Scientist*, p. 10.

24   Ibid., pp. 16–17.

25   As part of the war effort, the Unionist Government created a series of wartime production boards beginning in 1917. Finally, to placate farmers, the government created the Canadian Wheat Board in 1919, with a mandate to last one year. Ottawa did not renew its legislation the following year, dissolving the Canadian Wheat Board in 1920. T.C. Byrne, *Alberta's Revolutionary Leaders* (Calgary: Detselig Publishers, 1991), p. 37.

26   In the aftermath of the war, Alberta, like the rest of Canada, teemed in labour unrest. In the background, the IWW directed efforts through a new creation, the OBU - One Big Union. The OBU successfully managed massive strikes across the prairies in 1919 and even published a paper, *The Soviet*, in Edmonton. James G. MacGregor, A History of Alberta (Edmonton: Hurtig Publishers, 1981), pp. 239–40.

27   RG 33/52, "History of Negotiations re Transfer of the Natural Resources of Manitoba, Saskatchewan and Alberta," in Breen, ed., Select Documents, vol.1 of 4 vols.

28   RG 33/52, Letter, J.B.M. Baxter to W.F.A. Turgeon, February 23, 1929, in Breen, ed., *Select Documents*, vol. 1 of 4 vols.

29   Breen, "Canadian Petroleum Policy," pp. 292–93.

30   Kenneth Norrie and Douglas Owram, *A History of the Canadian Economy* (Toronto: Harcourt Brace & Co., 1996), pp. 320–24.

31   Incorporated in 1880, Imperial Oil was the largest Canadian oil company before Standard Oil bought a majority interest (69%) in the company in 1898 for $350,000. Imperial remained a subsidiary of

Standard Oil of New Jersey until 1911 when the Supreme Court broke the trust. Freehold lands were those lands in which individuals or large companies acquired mineral and surface rights. The practice was common in the late 19th century as the Canadian Pacific Railway distributed large tracts of freehold land in Western Canada to encourage settlement of the prairies. Depending on the jurisdiction, the provincial or federal government regulates the exploitation of resources, but permits private sales of oil and gas rights on freehold lands. George S. Gibb and Evelyn H. Knowlton, *History of Standard Oil Company (of New Jersey): The Resurgent Years, 1911–1927* (New York: Harper, 1956), pp. 260, 275.

32 Daniel Yergin, *The Prize: The Epic Quest for Oil, Money and Power* (New York: Simon & Schuster, 1992), pp. 189–90.

33 Gerald D. Nash, *United States Oil Policy: 1890–1964* (Pittsburgh: University of Pittsburgh Press, 1968), pp. 49–56, 60–65.

34 Breen, "Canadian Oil Policy," pp. 300–303.

35 Sheppard, *Oil Sands Scientist*, p. 17.

36 Barry Glen Ferguson, *The Athabasca Oil Sands: Northern Resource Exploration, 1875–1951* (Edmonton: Alberta Culture, 1985), p. 33.

37 It may have been yet another coincidence, but the student in question, W.F. Seyer, worked under the direction of a member of the Honorary Advisory Council before attending McGill University.

38 Ferguson, *Athabasca Oil Sands*, p. 28.

39 Ibid., p. 29.

40 The process of "destructive distillation" is a very basic form of separating substances. Heat is added to a compound, in this instance, tar sand, until it begins to break down into its constituent elements.

41 Letter, Clark to Dr. H.M. Tory, October 27, 1920, Document #2 in Sheppard, *Oil Sands Scientist*, p. 107.

42 Ibid.

43 Letter, Clark to Dr. H.M. Tory, November 18, 1921, Document #7 in Sheppard, *Oil Sands Scientist*, pp. 116–21.

44 Ferguson, *Athabasca Oil Sands*, pp. 35–37.

45 Yergin, *The Prize*, p. 208.

46 Beaver, "The U.S. Failure to Develop Synthetic Fuels," p. 245.

47 Ibid., p. 243. Emphasis mine.

48 Nash, *United States Oil Policy*, p. 49; Yergin, *The Prize*, pp. 222–23.

49 S.C. Ells, "Bituminous Sands of Northern Alberta," Mines Branch, *Department of Mines Report #632* (Ottawa: King's Printer, 1926), p. 101.

50 Letter, Dr. Clark to Blair, November 19, 1927, UAA SMB File 46/2/2/1/2.

51 Karl A. Clark, "Bituminous Sand Development," *The Press Bulletin*, February 15, 1929.

52　Jack L. Granatstein, et al, *Twentieth Century Canada* (Toronto: McGraw-Hill, Ryerson, 1986), 2d ed., pp. 197, 203.

53　MacGregor, *A History of Alberta*, p. 247.

54　Letter, Clark to Blair, November 19, 1927, UAA SMB, File 46/2/2/1/2.

55　Clark, "Memo: Re Bituminous Sand Studies During 1929–1930," March 26, 1929, Document #27 in Sheppard, *Oil Sands Scientist*, pp. 167–70.

56　Since 1921 James M. McClave, a hydro-metallurgist from Denver, had worked on a different hot water separation process. The chief differences between the two methods were that the McClave process used bentonite as a treating agent and the McClave process deliberately introduced additional air into the flotation cell to bring the oil to the surface for skimming.

57　J. Joseph Fitzgerald, *Black Gold With Grit* (Sidney: Gray's Publishing, 1978), p. 69.

58　Clark Field Diary entry, May 26, 1929, UAA Sidney Martin Blair Papers, File 46/2/2/1/3.

59　Letter, Clark to Dr. HM. Tory, President, National Research Council, Ottawa, December 1929, Document #30 in Sheppard, *Oil Sands Scientist*, pp. 176–96.

60　Clark, "Memo Re Bituminous Sand Studies During 1929–1930," March 26, 1929, Document #27 in Sheppard, *Oil Sands Scientist*, pp. 167–70.

61　Letter, Clark to Blair, December 27, 1930, UAA SMB File 46/2/2/1/3.

62　Ibid.

63　Letter, Clark to Dr. R.C. Wallace, March 4, 1930, Document # 32 in Sheppard, *Oil Sands Scientist*, pp. 197–99. Ferguson, Athabasca Oil Sands, p. 87.

64　Ibid.

65　Fitzgerald, *Black Gold With Grit*, p. 70.

66　Comfort, *The Abasand Fiasco*, p. 48.

67　Ibid.

68　Ferguson, *Athabasca Oil Sands*, pp. 85–95.

Chapter 2

1　Comfort, *Abasand Fiasco*, p. 92.

2　*Alberta Scrapbook Hansard, 1944*, pp. 103–4.

3　Byrne, *Alberta's Revolutionary Leaders*, p. 144.

4　Robert Bothwell and William Kilbourn, *C.D. Howe: A Biography* (Toronto: McClelland & Stewart, 1979), pp. 134–35.

5　De J. Kennedy, *History of the Department of Munitions and Supply: Canada in the Second World War*, vol. 2 (Ottawa: King's Printer, 1950), vol. 1, p. v.

6   John G. Diefenbaker, *One Canada* (Toronto: Macmillan, 1975), vol. 1, p. 240.

7   Alan S. Milward, *War, Economy and Society, 1939–1945* (Berkeley: University of California Press, 1977); Richard Overy, *Why the Allies Won*, p. 233; Harold F. Williamson, et al., *The American Petroleum Industry: The Age of Energy, 1899–1959* (Evanston: Northwestern University Press, 1963), vol. 2, pp. 747–49, 762.

8   Peter MacKenzie-Brown, Gordon Jaremko, and David Finch, *The Great Oil Age: The Petroleum Industry in Canada* (Calgary: Detselig Publishers, 1993), p. 15.

9   C.D. Howe, *Hansard*, vol. 4, November 7, 1941, pp. 4186–87.

10   Bothwell and Kilbourn, *C.D. Howe*, pp. 150–53.

11   Norman Hillmer and Jack L. Granatstein, *Empire to Umpire: Canada and the World to the 1990's* (Toronto: Copp Clark Longman, 1994), pp. 161–63.

12   Kennedy, *History of the Department of Munitions and Supply*, p. 153.

13   Larry Kohler, "Canadian-American Oil Diplomacy: The Adjustment of Conflicting National Oil Policies, 1955–1973" (Unpublished Ph.D. Dissertation, The Johns Hopkins University, 1983), p. 31.

14   John W. Frey and H.C. Ide, eds., *A History of the Petroleum Administration For War: 1941–1945* (Washington: Government Printing Office, 1946), p. 24.

15   Williamson, *The American Petroleum Industry*, pp. 775–82.

16   Yergin, *The Prize*, p. 395.

17   Breen, *Alberta's Petroleum Industry*, p. 192.

18   Ferguson, *Athabasca Oil Sands*, p. 101.

19   Ells, "Recollections," pp. 87–95.

20   Ibid., p. 102.

21   Cora Casselman, *Hansard*, vol. 1, p. 429.

22   C.D. Howe, *Hansard*, vol. 1. February 18, 1942, p. 686.

23   John R. MacNicol, *Hansard*, vol. 3. May 15, 1942, pp. 2472–78.

24   Ibid., p. 2478.

25   Charles Johnson, *Hansard*, vol. 3. May 19, 1942, pp. 2544–47.

26   Ibid., pp. 2562–63.

27   The Canol Oil Project in Norman Wells was supposed to supply oil product along the route of the proposed Alaskan highway and high-octane aviation fuel for planes based in Alaska. Under the terms of the agreement, the United States would bear the costs of building and operating the pipeline and refinery. At the end of the war, the Canadian government had first right of refusal to purchase the facility. The project ended up causing hard feelings on both sides of the border. Many Americans believed that the terms of the agreement were far too favourable to Canada while Canadians chafed at real or perceived ex-

cesses by the United States Army, particularly unconfirmed reports that the U.S. Army would answer the telephone with the phrase "American Army of Occupation." See Randall, *United States Foreign Oil Policy*, pp. 160–65; John H. Thompson and Stephen Randall, *Canada and the United States: Ambivalent Allies* (Montreal: McGill-Queen's University Press, 1994), pp. 168–69; Jack L. Granatstein and Norman Hillmer, For Better or For Worse: Canada and the United States to the 1990's (Toronto: Copp Clark Pittman, 1991), p. 154.

28   K.A. Clark, "Athabasca Oil Sands: Historical Review and Summary of Technical Data," *Research Council of Alberta*, Contribution #69 (Edmonton: Queen's Printer, 1957), p. 7.

29   Letter, C.D. Howe to Premier William Aberhart, June 23, 1942, *Alberta Scrapbook Hansard, 1944*, p. 103.

30   Ibid.

31   Letter, Aberhart to Howe, July 3, 1942, *Alberta Scrapbook Hansard, 1944*, p. 103.

32   On the Manhattan Project see Richard Rhodes, *The Making of the Atomic Bomb* (New York: Simon & Schuster, 1986).

33   Letter, Aberhart to Howe, July 3, 1942, *Alberta Scrapbook Hansard, 1944*, p. 103.

34   Ibid.

35   Ferguson, *Athabasca Oil Sands*, p.109.

36   Ibid., p. 124.

37   Ibid., p. 104.

38   Sheppard, *Oil Sands Scientist*, p. 71

39   Ferguson, *Athabasca Oil Sands*, p. 104.

40   Karl A. Clark, "Introduction to Consolidated Mining and Smelting Company Limited Report on Blasting Oil Sands at Abasand Oils, Ltd.," September 12–19, 1942, UAA SMB, File 46/2/2/1/3.

41   Ferguson, *Athabasca Oil Sands*, p. 102.

42   Ibid., p. 109.

43   Canada Department of Mines and Resources, "Drilling and Sampling of Bituminous Sands of Northern Alberta," vol. 1 of 3 vols. (Ottawa: King's Printer, 1949), p. 1.

44   Comfort, *The Abasand Fiasco*, pp. 76–82.

45   C.D. Howe, *Hansard*, vol. 3, April 14, 1943, pp. 2152–53.

46   Ells, "Recollections," p. 93.

47   Ferguson, *Athabasca Oil Sands*, p. 111.

48   C.D. Howe, *Hansard*, vol. 3, April 14, 1943, p. 2152.

49   Fitzgerald, *Black Gold With Grit*, p. 83.

50   Ibid.

51   C.D. Howe, *Hansard*, vol. 4, June 14, 1943, pp. 3625–26.

52 One observer pointed out that Howe accepted Calvin Coolidge's asser-tion that "the business of America is business" and added the corollary "and the sooner Canada lives by that rule, the better off we'll be." Christina McCall-Newman, *Grits: An Intimate Portrait of the Liberal Party* (Toronto: Macmillan, 1982), p. 26.

53 Letter, Karl Clark to Earle Smith, General Manager, Abasand Oils Ltd., May 21, 1943, Document #79 in Sheppard, *Oil Sands Scientist*, pp. 317–20.

54 Letter, Clark to E.O. Lilge, Consolidated Mining and Smelting Company, July 22, 1943, Document #80 in Sheppard, *Oil Sands Scientist*, pp. 320–21.

55 Ibid.

56 Comfort, *The Abasand Fiasco*, p. 82; Letter, Clark to G.D. Raitt, July 23, 1943, Document #81 in Sheppard, p. 322. John R. MacNicol, *Hansard*, vol. 4, June 14, 1943, pp. 3626–27.

57 Ferguson, *Athabasca Oil Sands*, p. 112.

58 Ibid., p. 113.

59 Memorandum, "Interview with Mr. P.D. Hamilton of General Engineering Co.," Document #83 in Sheppard, *Oil Sands Scientist*, pp. 324–26.

60 Ibid.

61 Ferguson, *Athabasca Oil Sands*, p. 117.

62 Letter, Premier Manning to C.D. Howe, January 7, 1944, *Alberta Scrapbook Hansard, 1944*, p. 104.

63 Ibid.

64 Ibid.

65 Letter, C.D. Howe to Premier Manning, January 22, 1944, *Alberta Scrapbook Hansard, 1944*, p. 104.

66 Ibid.

67 Ibid. Emphasis mine.

68 Ibid.

69 Clark to Hon. W.A. Fallow, February 29, 1944, Letter #86 in Sheppard, *Oil Sands Scientist*, pp. 328–37.

70 Ibid.

71 Ibid.

72 *Alberta Scrapbook Hansard, 1944*, March 13, 1944, pp. 103–4.

73 Ibid.

74 C.D. Howe, *Hansard*, vol. 2, March 21, 1944, p. 1722.

75 Ibid.

76 Ibid.

77 Ibid.

78 Ibid., p. 1737.

79 Ferguson, *Athabasca Oil Sands*, pp. 117–18.

80 Thomas Crerar, *Hansard*, vol. 4, May 23, 1944, pp. 3211–13.

81 Ibid., p. 3213.

82 Ibid., p. 3219.

83 Ibid., pp. 3221–31.

84 Ibid., pp. 3232–33.

85 Ibid. p. 3235.

86 Ibid., p. 3241.

87 *Alberta Scrapbook Hansard 1944*, March 14, 1944, p. 75.

88 J.A. Glen, *Hansard*, vol. 3, December 3, 1945, pp. 2864–81.

89 Fitzgerald, *Black Gold With Grit*, p. 85.

90 Yergin, *The Prize*, p. 369.

Chapter 3

1 Letter, Karl A. Clark to Ray D. Magladry, Ottawa, July 8, 1947, Document #124 in Sheppard, *Oil Sands Scientist*, pp. 398–400.

2 Michael B. Stoff, *Oil, War, and American Security: The Search for a National Policy on Foreign Oil, 1941–1947* (New Haven: Yale University Press, 1980), p. 70.

3 *Alberta Scrapbook Hansard, 1944*, March 13, 1944, p. 105.

4 Ferguson, *Athabasca Oil Sands*, p. 126.

5 *Alberta Scrapbook Hansard, 1945*, March 1, 1945, p.15.

6 Thompson and Randall, *Ambivalent Allies*, p. 179.

7 Eric Sievwright, "The Effect of Petroleum Development on the Alberta Economy, 1947–1957," (Unpublished Ph.D. Dissertation. McGill University, 1960), p. 78.

8 Kohler, "Canadian-American Oil Diplomacy," pp. 8, 28–29.

9 Kenneth Norrie and Douglas Owram, *A History of the Canadian Economy* (Toronto: Harcourt Brace & Co., 1996), p. 411.

10 Robert Bothwell, Ian Drummond, and John English, *Canada Since 1945* (Toronto: University of Toronto Press, 1989), p. 142, rev. ed.

11 Robert Bothwell, *Canada and the United States: The Politics of* Partnership (Toronto: University of Toronto Press, 1992), p. 28.

12 Hillmer and Granatstein, *Empire to Umpire*, p. 194.

13 Ibid.

14 Robert Cuff and J.L. Granatstein, "The Rise and Fall of Canadian-American Free Trade, 1947–1948," *Canadian Historical Review* 57 (4), pp. 469, 473; Donald Creighton, *Canada's First Century* (Toronto: Macmillan, 1970), pp. 242–43; Granatstein and Hillmer, *For Better or For Worse*, pp. 170–75; Hillmer and Granatstein, *Empire to Umpire*, pp.196–99.

15 Granatstein and Hillmer, *For Better or For Worse*, p. 168.

16   Data in table taken from Breen, *Alberta's Petroleum Industry*, p. 559.

17   Ibid., p. 248.

18   Smith, *Treasure Seekers*, pp. 102–3.

19   Canada Department of Mines and Resources, "Drilling and Sampling of Bituminous Sands of Northern Alberta," vol. 1. *Results of Investigations, 1942–1947* (Ottawa: Government Printer, 1949).

20   Ferguson, *Athabasca Oil Sands*, p. 124.

21   Ibid., pp. 124–26.

22   Ibid., p. 126.

23   Ibid.

24   *Alberta Scrapbook Hansard, 1945*, March 22, 1945, p. 79.

25   Ibid.

26   Letter, Clark to Dr. Sidney Born, Born Engineering Company, Tulsa, Oklahoma. June 20, 1945, Document #107 in Sheppard, *Oil Sands Scientist*, p. 370.

27   Ibid.

28   Letter, Clark to C.J. Knighton, September 18, 1945, Document #111 in Sheppard, *Oil Sands Scientist*, pp. 374–75.

29   Letter, K.A. Clark to S.M. Blair, July 20, 1946, UAA SMB File 46/2/2/1/12.

30   Ibid.

31   Ibid.

32   Nash, *United States Oil Policy*, p. 187.

33   Michael J. Hogan, *A Cross of Iron: Harry S. Truman and the Origins of the National Security State, 1945–1954* (New York: Cambridge University Press, 1998), pp. 209–17.

34   Halford L. Hoskins, *Middle East Oil in United States Foreign Policy* (Westport: Hyperion Press, 1950), p. 5

35   David S. Painter, *Oil and the American Century: The Political Economy of U.S. Foreign Oil Policy, 1941–1954* (Baltimore: The Johns Hopkins University Press, 1986), p. 99.

36   Melvyn P. Leffler, "National Security, Core Values, and Power," in Thomas G. Paterson and Dennis Merril, eds., *Major Problems in American Foreign Relations, Volume II: Since 1915.* 4th ed. (Lexington: D.C. Heath & Company, 1995), pp. 14–15; Thomas J. McCormick, *America's Half-Century: United States Foreign Policy in the Cold War and After*, 2d ed. (Baltimore: The Johns Hopkins University Press, 1995), pp. 4–5; Eckes, *Opening America's Market*, p. 157; Painter, *Oil and the American Century*, p. 99.

37   Hoskins, *Middle East Oil in United States Foreign Policy*, p. 25

38   Richard H.K. Vietor, "The Synthetic Liquid Fuels Program: Energy Politics in the Truman Era," *Business History Review* LVI (1980), pp. 24–25.

39   Hoskins, *Middle East Oil in United States Foreign Policy*, p. 25.

40   Ibid., pp. 17–28; Vietor, "The Synthetic Liquid Fuels Program," p. 19.

41   Nash, *United States Oil Policy*, pp. 210–11.

42   For a discussion of the early relationship between the United States government and American oil companies operating in Saudi Arabia, see Stoff, *Oil, War, and American Security*, pp. 34–61.

43   Leffler, *Preponderance of Power*, pp. 423, 497.

44   Report to the National Security Council, December 28, 1950, *FRUS, 1950*, Vol. 1, pp. 489–49; Acting Secretary of Defense to Executive Secretary of the National Security Council, December 27, 1950, *FRUS, 1950*, Vol. 1, pp. 490-492; Draft Letter from Executive Secretary of the National Security Council to the Secretary of the Interior, December 28,1950, *FRUS, 1950*, Vol. I, p. 492.

45   Leffler, *A Preponderance of Power*, pp. 77–80.

46   Yergin, *The Prize*, p. 410; James Bamberg, *British Petroleum and Global Oil: The Challenge of Nationalism* (Cambridge: Cambridge University Press, 2000), p. 10.

47   Breen, *Alberta's Petroleum Industry*, pp. 245–46.

48   Ibid., p. 246.

49   Sievwright, "The Effect of Petroleum Development on the Alberta Economy," p. 62.

50   Letter, Clark to Ray D. Magladry, Ottawa, July 8, 1947, Document #124 in Sheppard, *Oil Sands Scientist*, pp. 398–400.

51   Since the Canadian oil industry seemed to be "Canadian" in name only, it must be considered as part of an integrated North American market even as early as 1947. American multinationals made many of the decisions affecting the operations of their subsidiaries and now had to integrate this new production into the North American market. With comparatively small amounts of oil being produced in Canada, however, many observers thought it unlikely that Canada could furnish the United States with significant amounts of crude oil. Nevertheless, Canadian oil did manage to service particular markets in portions of the United States precisely because a large proportion of American multinational companies developed Canada's comparatively meagre oil reserves. Canadian politicians increasingly allowed their national oil industry to develop along "continental" lines by servicing particular markets in the United States, rather than developing a national solution – that is, developing an Eastern Canadian market for Alberta's oil.

52   Yergin, *The Prize*, pp. 499–500.

53   Ibid.

54   Peter Grose, *Gentleman Spy: The Life of Allen Dulles* (New York: Houghton Mifflin, 1994), p. 46.

55   Of course, Leith was not the only advocate of conserving domestic American crude supplies while importing oil from foreign sources. The idea received careful consideration in the State Department as early as November 1941. See Emily Rosenberg, *Spreading the American Dream: American Economic and Cultural Expansion, 1890–1945* (New York: Hill & Wang, 1982), pp. 126–27, 196–97; Stoff, *Oil, War, and American Security*, p. 67; John L. Gaddis, *We Now Know: Rethinking Cold War History* (New York: Oxford University Press, 1997), p. 196; Alfred E. Eckes, Jr., "U.S. Trade History" in William A. Lovett, Alfred E. Eckes, Jr., and Richard Brinkman, *U.S. Trade Policy: History, Theory, and the WTO* (New York: M.E. Sharpe, 1999); Kunz, *Butter and Guns*; Kohler, "Canadian-American Oil Diplomacy," p. 1.

56   Vietor, "The Synthetic Liquid Fuels Program," p. 15.

57   Ibid.

58   Ibid., p. 27.

59   Alberta Research Council, *Thirtieth Annual Report, 1949*, Report #55 (Edmonton: King's Printer, 1950), p. 8.

60   Ibid., pp. 9–10.

61   Letter, Clark to Sidney C. Ells, November 8, 1948, Document #150 in Sheppard, *Oil Sands Scientist*, pp. 438–42.

62   Ibid.

63   Oil Sands Project, "Report to the Board of Trustees of the Oil Sands Project From Inception to December 31, 1948," Sessional Paper #53 (Edmonton: King's Printer, 1949).

64   Byrne, *Alberta's Revolutionary Leaders*, p. 179.

65   Karl A. Clark, "Bituminous Sands Investigations, September 1949." (Edmonton: King's Printer, 1949).

66   Ibid.

67   Letter, Edwin Nelson to the Honourable John L. Robinson [Minister, Department of Industries and Labour], November 18, 1949; Letter, Honourable John L. Robinson to Edwin Nelson, November 22, 1949, UAA SMB File 46/2/2/2/15.

68   Letter, Clark to Malcolm Clark, September 25, 1949, Document #160 in Sheppard, *Oil Sands Scientist*, pp. 454–57.

Chapter 4

1   Board of Trustees, Oil Sands Project, *Proceedings: Athabasca Oil Sands Conference* (Edmonton: King's Printer, 1951), p. 167.

2   *Alberta Scrapbook Hansard, 1949*. March 22, 1949, p. 59.

3   K.A. Clark, "Athabasca Oil Sands: Historical Review and Summary of Technical Data." Alberta Research Council, Contribution #69 (Edmonton: Queen's Printer, 1957), p. 15.

4   Memorandum, Karl A. Clark, "Matters Needing Attention When the Bitumount Plant is Readied for Further Operation," January 1950, UAA SMB File 46/2/2/1/12.

5   Eric J. Hanson, *The Dynamic Decade: The Evolution and Effects of the Oil Industry in Alberta* (Toronto: McClelland and Stewart, 1958), pp. 176–77.

6   Canada, *Royal Commission on Energy, Second Report* (Ottawa: Government Printer, 1959), p. 13, hereafter cited as *Borden Commission Second Report*; Hanson, *Dynamic Decade*, pp. 201–3.

7   Alvin Finkel, *The Social Credit Phenomenon in Alberta* (Toronto: University of Toronto Press, 1989), p. 100.

8   Ferguson, *Athabasca Oil Sands*, p. 140.

9   Alberta Research Council, *Thirty-First Annual Report* (Edmonton: Government Printer, 1950) pp. 13–14.

10 Letter, W.E. Adkins to S.M. Blair, March 10, 1950, UAA SMB File 46/2/2/2/8.

11 Letter, S.M. Blair to W.E. Adkins, March 3, 1950, UAA SMB File 46/2/2/2/8.

12 Letter, W.E. Adkins to G.W. Hodgson, March 13, 1950, UAA SMB File 46/2/2/2/8.

13 Letter, S.M. Blair to W.E. Adkins, March 16, 1950, UAA SMB File 46/2/2/2/8.

14 Letter, W.E. Adkins to S.M Blair, March 18, 1950, UAA SMB File 46/2/2/2/8.

15 Letter, S.R. Blair to S.M. Blair, June 16, 1950, UAA SMB File 46/2/2/2/2.

16 Letter, S.M. Blair to the Honourable John L. Robinson, October 18, 1950, UAA SMB File 46/2/2/2/8.

17 Sidney M. Blair, *Report on the Alberta Bituminous Sands* (Edmonton: Government of Alberta, 1950), p. 7.

18 Ibid.

19 *Edmonton Bulletin*, January 6, 1951.

20 Andrew Snaddon, "Alberta Politics," *The Calgary Herald*, January 6, 1951.

21 The Manning government promised it would not export significant amounts of petroleum until the province had proven reserves to satisfy Alberta consumption for thirty years. Breen, *Alberta's Petroleum Industry and the Conservation Board*, p. 258.

22 "Oil Could be 'Mined' from Sands at Profit," *The Calgary Herald*, May 16, 1951.

23 "U.K. Firm's Experts to Study Development of Oil Sands," *Edmonton Journal*, March 19, 1951.

24 "'Big Look' at McMurray," *Edmonton Journal*, March 20, 1951.

25 *Edmonton Journal*, May 25, 1951.

26 Breen, *Alberta's Petroleum Industry*, p. 444.

27 Letter, John Robinson to S.M. Blair, August 11, 1951, UAA SMB File 46/2/2/3/3.

28 Honourable N.E. Tanner, "Government Policy Regarding Oil-Sand Leases and Royalties," *Proceedings: Athabasca Oil Sands Conference* (Edmonton: King's Printer, 1951), p. 176.

29 Ibid.

30 Ibid., p. 177. Emphasis in original.

31 Ibid.

32 K.A. Clark, "Athabasca Oil Sands," pp. 18–19.

33 Breen, *Alberta's Petroleum Industry and the Conservation Board*, p. 445.

34 Nash, *United States Oil Policy*, pp. 202–6. Robert L. Bradley, Jr., *Oil, Gas and Government*, Vol. 1 (London: Rowman & Littlefield Publishers, 1996), pp. 726–29.

35 Douglas Bohi and Milton Russell, *Limiting Oil Imports* (Baltimore: Johns Hopkins University Press, 1978), p. 28; Bradley, *Oil, Gas and Government*, p. 730; Yergin, *The Prize*, pp. 535–36.

36 Kunz, *Butter and Guns*, pp. 225–26.

37 Breen, *Alberta's Petroleum Industry*, p. 445.

38 Great Canadian Oil Sands was the product of a deal between Abasand Oils Limited, Canadian Oil Companies Limited and Sun Oil Company of Philadelphia.

39 *Borden Commission Second Report*, p. 126.

40 Kohler, "Canadian-American Oil Diplomacy," p. 30.

41 Breen, *Alberta's Petroleum Industry and the Conservation Board*, p. 446.

42 Nash, *United States Oil Policy*, p. 205.

43 Kohler, "Canadian-American Oil Diplomacy," p. 74.

44 Theodore Binnema, "Making Way for Canadian Oil: UnitedStates Policy, 1947–1959," *Alberta History* 45 (1997), p.19. For two different opinions on the impact of the voluntary quota program compare Ed Shaffer, *Canada's Oil and the American Empire* (Edmonton: Hurtig, 1983), p. 148; and Alan R. Plotnick, *Petroleum: Canadian Markets and United States Foreign Trade Policy* (Seattle: University of Washington Press, 1964) p. 114–15, 117.

45 Breen, *Alberta's Petroleum Industry*, p. 458.

46 Canada, *Royal Commission on Energy, First Report* (Ottawa: Government Printer, 1958), p. 90.

47 The construction of a trans-Canadian pipeline was a very sensitive political matter that had already cost the Liberal Party dearly. In 1957, the government of Louis St. Laurent invoked early closure on a debate about providing loans to an American-owned company to build a natural gas pipeline from Alberta to southern Ontario. The closure set off a

veritable firestorm of protest. "From early May to 'Black Friday,' June 1, the House was in a tragi-comic chaos, with sessions ending at dawn and animal noises echoing throughout the chamber," wrote Lester Pearson's biographer John English. "On Black Friday, the speaker declared it was Thursday, the centre aisle swelled with angry members, and the Chamber echoed with 'Onward Christian Soldiers' from Tories and the CCF, while the Liberals tried to drown them out with 'We've been Working on the Pipeline.' It was, perhaps, one of Parliament's blackest moments." Labelled the "American pipeline buccaneers" by John Diefenbaker in the subsequent election, the Conservative Party scored a major upset victory over the Liberals in 1957. So certain were many pundits about the outcome, *Maclean's* magazine went to press with an editorial that began, "For better or for worse, we Canadians have once more elected one of the most powerful governments ever created." Diefenbaker would end up consolidating his position in March 1958 with a solid victory at the polls. John English, *The Worldly Years* (Toronto: Knopf Canada, 1992) p. 157; Binnema, "Making Way for Canadian Oil," p. 20; William Kilbourn, *Pipeline: TransCanada and the Great Debate* (Toronto: Clarke Irwin, 1970).

48   Breen, *Alberta's Petroleum Industry*, p. 459.

49   *Borden Commission Second Report*, p. 113.

50   However, Imperial operated the largest refinery in the Montreal market and, at 70,000 barrels per day, imported the most amount of duty-free Venezuelan crude. Breen, *Alberta's Petroleum Industry*, p. 438.

51   *Borden Commission Second Report*, pp. 109, 130.

52   Alberta Research Council, "Submission to Borden Commission," p. 21, PAA File 68.15/49, Box 2.

53   Ibid., p. 22.

54   Alberta Technical Committee "Report to the Minister of Mines and Minerals and the Oil and Gas Conservation Board: With Respect to an Experiment Proposed by Richfield Oil Corporation Involving an Underground Nuclear Explosion Beneath the McMurray Oil Sands with the Objective of Determining the Feasibility of Recovering the Oil with the Aid of the Heat Released from such an Explosion" (Edmonton: Government Printer, 1959).

55   Breen, *Alberta's Petroleum Industry*, p. 450.

56   Clark, "Athabasca Oil Sands: Historical Review and Summary of Technical Data," pp. 5–6.

57   Breen, *Alberta's Petroleum Industry*, p. 448. Emphasis mine.

58   *Borden Commission Second Report*, p. 126.

59   Ibid., p. 144.

60   G. Bruce Doern and Glen Toner, *The Politics of Energy: The Development and Implementation for the NEP* (Toronto: Methuen, 1984), pp. 80–81.

61 Breen, *Alberta's Petroleum Industry*, pp. 498–99.

62 Binnema, "Making Way for Canadian Oil," p. 22; Yergin, *The Prize*, p. 513.

63 NSC5822/1 National Security Council, "Certain Aspects of U.S. Relations with Canada," December 30, 1958, p. 2. [Online]: <http: //nsarchive.chadwyck.com> [15 February 2002].

64 Ibid., p. 5.

65 Arthur M. Johnson, *The Challenge of Change: The Sun Oil Company, 1945–1977* (Columbus: Ohio State University Press, 1983), p. 101.

Chapter 5

1 Graham D. Taylor, "Sun Oil Company and Great Canadian Oil Sands Ltd.: The Financing and Management of a 'Pioneer' Enterprise, 1962–1974," *Journal of Canadian Studies* 20 (1985), p. 104.

2 Editorial, *Oil and Gas Journal*, October 23, 1967, p. 39.

3 Bill Tisdall, "Exploration and Production Activities at GCOS," p. 1, unpublished manuscript, GA M8057 Ned Gilbert Fonds.

4 George Dunlap, "Miscellaneous," p. 13, unpublished manuscript, GA M8057 Ned Gilbert Fonds.

5 Ibid.

6 See Chapter 4.

7 Tisdall, "Exploration and Production Activities at GCOS," p. 2.

8 Ned Gilbert, "The Suncor Story," p. 9, unpublished manuscript, GA M8057 Ned Gilbert Fonds.

9 Ibid.

10 Johnson, *Challenge of Change*, p. 128.

11 Taylor, "Sun Oil Company and the Great Canadian Oil Sands Ltd.," pp. 108–9; Johnson, *Challenge of Change*, p. 128.

12 James D. Hilborn, "After Leduc," in Hilborn, ed., *Dusters and Gushers: The Canadian Oil and Gas Industry*, p. 198; *Borden Commission Second Report*, p. 172.

13 *Borden Commission Second Report*, p. 11.

14 Kohler, "Canadian-American Oil Diplomacy," p. 195.

15 *Calgary Herald*, February 23, 1960.

16 The Progressive Conservatives, led by Peter Lougheed, in 1971 replaced the Social Credit government. Since 1971, the Progressive Conservatives have ruled provincial politics.

17 Brooke Jeffrey, *Hard Right Turn: The New Face of Neo-Conservatism in Canada* (Toronto: HarperCollins, 1999), p. 54.

18 *Scrapbook Hansard, 1960*, March 14, 1960, p. 77.

19 Oil and Gas Conservation Board, *Report to the Lieutenant Governor in Council With Respect to the Application of Great Canadian Oil Sands*

Limited Under Part VIA of the Oil and Gas Conservation Act (Edmonton: Government Printer, 1960). Hereafter cited as OGCB 60-6.

20   Historian David Breen called the Conservation Board "one of the most important regulatory bodies in Canada" because the impact of its decisions extended well beyond Alberta's borders. Established in 1938 to protect the long-term interest of Albertans in the province's natural resources, the Oil and Gas Conservation Board was an independent arm of the provincial government. Staffed by industry professionals, the size of the Board expanded as the petroleum industry grew. In September 1948, the Board employed 31 people in a variety of positions ranging from engineers to office staff. Ten years later, the Board's staff had increased to 206. Breen, *Alberta's Petroleum Industry and the Conservation Board*, p. xvii. OGCB 60–6, p. 1.

21   Ibid., p. 3.

22   Ibid., pp. 73–75.

23   Ibid., pp. 80–81.

24   Frank Dabbs, "How it All Began: IPAC at 30," *Oilweek*, January 29, 1990, p. 10.

25   *Alberta Scrapbook Hansard, 1962*, February 24, 1962, p. 18.

26   *Alberta Scrapbook Hansard, 1962*, March 15, 1962, p. 59; March 22, 1962, p. 76A.

27   OGCB 62-7, p. 39.

28   Ibid., pp. 32, 45, 46.

29   Sun Oil's involvement in the oil sands has been the subject of a handful of scholarly monographs that benefited greatly from access to the company's private records. August W. Giebelhaus, *Business and Government in the Oil Industry: A Case Study of Sun Oil, 1876–1945* (Greenwich: Jai Press Inc., 1980); Johnson, *Challenge of Change*.

30   Johnson, *Challenge of Change*, p. 142.

31   Ibid., p. 9.

32   Finkel, *Social Credit Phenomenon*, pp. 141–76. Johnson, *Challenge of Change*, p. 129. Frank Dabbs, "Oil Sands Projects Stuck Between Past and Future," *Financial Post*, January 4, 1990, p. 14.

33   Text of the provincial government's oil sands policy available in Oil and Gas Conservation Board 63-8, "Government Policy Statement With Respect to Oil Sands Development," *Report to the Lieutenant Governor in Council With Respect to the Applications of Cities Service Athabasca Inc. and Shell* (Edmonton: Government Printer, 1963), pp. A1-A4. Hereafter cited as OGCB 63-8. The Board of Directors of the Canadian Petroleum Association decided to remain silent on the amended GCOS application. Minutes, CPA Board of Directors, December 13, 1962, GA CPA Box 3, File 18.

34   *Oilweek*, October 29, 1962, p. 15.

35   Ibid.

36   OGCB 63-8, p. 231.

37   Ibid., p. 232.

38   Ibid., p. 237. OGCB 62-7, p. 45.

39   *Alberta Scrapbook Hansard, 1962*, March 22, 1962, p. 89.

40   *Revised Statutes of Alberta*, Mines and Minerals Act, 1962, Part VI, Section 174, p. 197.

41   *Alberta Scrapbook Hansard, 1963*, March 15, 1963, p. 86.

42   Taylor, "Sun Oil and the Great Canadian Oil Sands Ltd.," pp. 108–9.

43   Taylor, "Sun Oil and the Great Canadian Oil Sands Ltd.," p. 109. "Athabasca Oil Sand Project Rumoured on Financial Rocks," *The Calgary Albertan*, September 28, 1963, p. 3; "Sun has 67 Million Cash to Back Oil Sand Project," *Oilweek*, October 7, 1963, p. 14.

44   Taylor, "Sun Oil and the Great Canadian Oil Sands Ltd.," p. 104.

45   OGCB 64-3, p. 58.

46   OGCB 64-3 p. 13.

47   Ibid., p. 36.

48   Ibid., pp. 73–74.

49   Ibid., pp. 59–60.

50   Ibid.

51   Ibid., p. 51.

52   Ibid., p. 75.

53   *Oilweek*, February 24, 1964, p. 35.

54   OGCB 64-3, p. 76.

55   "U.S. Shale Oil and Athabasca Oil May Hit Markets About Same Time," *Oilweek*, May 4, 1964, pp. 12–13, 28.

56   *Calgary Herald*, November 28, 1964, p. 52.

57   "Rapid Development Urged on Oil Sands," *Calgary Herald*, March 3, 1966, p. 26.

58   GCOS Progress Report, October 1965, p. 1.

59   *Calgary Albertan*, March 8, 1966, p. 24.

60   *Financial Post*, December 31, 1966, p. 3.

61   Syncrude Canada, "Submission Regarding Oil Sands Development Policy," May 11, 1966, p. 2. GA M6856 CPA Box 4, File 27.

62   Ibid., p. 4.

63   Report, "Oil Sands Development Policy May 11 Industry Meeting," George Govier, Chairman Oil and Gas Conservation Board, to Premier Manning, September 23, 1966, p. 6, GA M6856 CPA Box 4, File 27.

64   Ibid., p. 7.

65   Ibid., p. 17.

66   Ibid., pp. 3–4; Oil and Gas Conservation Board 68-C, "Report of An Application of Atlantic Richfield Co., Cities Service Athabasca, Inc., Imperial Oil, and Royalite Company Limited Under Part VI A of the

Oil and Gas Conservation Act," (Calgary: Government Printer, 1968), p. 69, hereafter cited as OGCB 6b-C; George Govier, "Alberta's Synthetic Crude Oil Development Policy," *Quarterly of the Colorado School of Mines* 65 (1970), p. 229.

67   Govier specifically mentions altering the definitions of "within reach" and "beyond reach" markets to increase production from the oil sands. Alberta Oil and Gas Conservation Board Report, "Oil Sands Development Policy," (Calgary: Government Printer, 1966), pp. 20–26.

68   Despite the Japanese government's intense interest in the sands, the province did not sign an agreement, perhaps because selling the oil to Japan would require extensive upgrading, processing, and refining in Canada before shipping the bitumen across the Pacific Ocean. The advantage of catering to the American market was that it limited the number of refining steps Canadian producers had to undertake. *Calgary Herald*, May 14, 1966, p. 1; *Calgary Albertan*, May 20, 1966, p. 22; "Japanese Firms Want 'In,'" *Calgary Albertan*, May 30, 1966.

69   The difference in their respective constituencies accounts for the different positions of the two organizations. Up until the late 1960s, more than half of the Canadian Petroleum Association's membership came from four companies – Imperial Oil, Shell Canada, BP Canada, and Gulf Canada – and the CPA clearly acted as the "voice of the majors" in discussions with various levels of government. On the other hand, the Independent Petroleum Association of Canada began operations in 1961 to give independent producers a voice in provincial and federal politics.

70   IPAC Press release, March 7, 1968, p. 2, GA IPAC roll 19 [ff. 161.0].

71   "Oil Sands Development Policy," February 1968, p. A7-A8, text in OGCB 68-C, pp. A4-A15.

72   105,000 barrels per day plus the 45,000 barrels per day already allocated to Great Canadian Oil Sands.

73   Ibid., p. A9-A13.

74   Ibid., p. 89.

75   Given the composition of the CPA – more than half its membership was derived from four large firms, two of whom had more than a passing interest in the oil sands – it is not difficult to see why the Association refrained from issuing a statement on the new development policy. Instead, CPA heavyweights, like Texaco, Chevron, Mobil, Pan Am, Hudson's Bay, Fina, and Triad, joined with IPAC to oppose the government's new policy.

76   IPAC Press release, October 27, 1967, GA IPAC roll 19 [ff. 150.0].

77   This would set the total amount of production from the oil sands at 125,000 barrels per day.

78   Letter, Charles S. Lee to members of the National Oil Committee, July 5, 1968, GA IPAC roll 19 [ff. 154.0]

79   Author unknown. "Syncrude Application," p. 6, GA IPAC Roll 19 [ff.158.5]. The manner in which this particular document was microfilmed makes it impossible to accurately determine exactly who the author was. Furthermore, part of the terms for IPAC's donation to the Glenbow Archives is that after microfilming, either IPAC or the Glenbow destroyed the original documents, making it impossible to check against the original.

80   Ibid., p. 12, GA IPAC Roll 19 [ff.160.0].

81   IPAC Press release, July 5, 1968, p. 1, GA IPAC roll 19 [ff 153.6].

82   Ibid., pp. 3–4, GA IPAC roll 19 [ff 153.6-153.7].

83   Walter J. Levy & Associates "U.S. Import Policy: Implications for Canadian Export of Conventional and Tar Sand Oil," July 1968.

84   "Of particular importance, it seems to the Board, will be a consideration of North American defence and the necessity of continued close co-operation between the US and Canada in all matters having a bearing on the security of North America," OGCB 68-C, pp. 181–83.

85   Ibid., pp. 234–35.

86   Syncrude Canada, "Submission Re: Syncrude Canada Ltd. to the Lieutenant Governor in Council, The Government of Alberta," 1969, text in Oil and Gas Conservation Board 69-B, "Supplement to the Report of Report of An Application of Atlantic Richfield Co., Cities Service Athabasca, Inc., Imperial Oil, and Royalite Company Limited under Part VI A of the Oil and Gas Conservation Act" (Calgary: Government Printer, 1969), pp. B1-B4.

87   Ibid., p. B2.

88   Ibid., p. 93.

89   Ibid., pp. 91–92.

Chapter 6

1   Alan Hustak, *Peter Lougheed* (Toronto: McClelland & Stewart, 1979), p. 162.

2   Canadian Petroleum Association, "Oil Sands Development – Risks, Economics Outlook," GA CPA Box 45, File 700.

3   Richard K. Vietor, *Energy Policy in America since 1945: A Study of Business-Government Relations* (Cambridge: Cambridge University Press, 1984), p. 329.

4   Anthony Cave Brown, *Oil, God, and Gold: The Story of Aramco and the Saudi Kings* (Boston: Houghton Mifflin, 1999), p. 299.

5   Yergin, *The Prize*, p. 615.

6 By way of comparison, $240 million represented nearly three-quarters of Canada's investment in the Saint Lawrence Seaway. Johnson, *The Challenge of Change*, pp. 138, 141.

7 The ceremonial opening of the GCOS plant took place in September 1967 even though the plant would not be operational until September 1968.

8 All information in this table provided by GCOS Annual Reports for stated year. Please note that totals in Profits (Losses) Column do not include additional construction costs of $69.9 million.

9 Johnson, *Challenge of Change*, p. 144.

10 Chas. Pearse, "Athabasca Tar Sands," *Canadian Geographical Journal* 76, p. 7.

11 Johnson, *Challenge of Change*, p. 158. Ernest Manning retired from politics in December 1968 and Harry Strom assumed the leadership of the Social Credit Party. A gentleman farmer, Strom was first elected to the Legislature in 1955 and served as minister of agriculture (1962–68). He briefly held the post of minister of municipal affairs (July-December, 1968) before becoming premier.

12 F.K Spragins, "Athabasca Tar Sands: Occurrence and Commercial Projects," GA M5873 Frank Spragins Fonds.

13 Ibid.

14 Stobaugh and Yergin, *Energy Future*, p. 3; Yergin, *The Prize*, pp. 567–68, 574–83.

15 Bothwell, *Canada and the United States*, p. 118.

16 Jack L. Granatstein and Robert Bothwell, *Pirouette: Pierre Trudeau and Canadian Foreign Policy* (Toronto: University of Toronto Press, 1990), p. 83.

17 Ibid.

18 Peter Foster, *Self Serve: How Petro-Canada Pumped Canadians Dry* (Toronto: Macfarlane & Ross, 1992), p. 55.

19 Canadian Petroleum Association "Submission to the Government of the Province of Alberta in the Matter of the Oil Sands Development Policy," November 1, 1971, p. 7, Appendix I, GA CPA Box 13, File 189.

20 Ibid., p. 7. Joe Greene later joked to the new Energy Minister, Donald Macdonald, "What did you do with all that oil I left you?" after the CPA's revised figures were announced. Foster, *Self-Serve*, p. 55.

21 Although Vernon originally applied his theory exclusively to developing countries, subsequent writers have expanded and modified the theory to apply to developed countries. See Raymond Vernon, *Sovereignty at Bay: The Spread of U.S. Enterprises* (New York: Basic, 1971), pp. 47–53; Robert Gilpin, *U.S. Power and the Multinational Corporation: The Political Economy of Foreign Direct Investment* (New York: Basic, 1975); Fred Bergsten, Thomas Horst, and Theodore Moran, *American*

Multinationals and American Interests (Washington: Brookings, 1978); John Richards and Larry Pratt, *Prairie Capitalism: Power and Influence in the New West* (Toronto: McClelland and Stewart, 1979); Barbara Jenkins, "Reexamining the 'Obsolescing Bargain': A Study of Canada's National Energy Program," *International Organization* 40 (1986), pp. 139–65.

22   An interesting development considering that the premier's brother, Don Lougheed, was an executive with Imperial Oil.

23   Peter Foster, *Blue-Eyed Sheiks: The Canadian Oil Establishment* (Toronto: Totem Books, 1979), p. 45.

24   Hustak, *Peter Lougheed*, p. 148.

25   The extent of these discussions is undetermined by reference to the documentary record because it was extremely rare to keep written records of this sort. However, there are enough indications in the Petroleum History's Oral History Project to suggest that Manning took an active interest in the health of the oil industry and smoothed matters over in cabinet when issues arose. GA M8615 Box 2, Ned Gilbert Fonds.

26   Hustak, *Peter Lougheed*, p. 140.

27   Ibid., p. 161, Pratt, *The Tar Sands*, pp. 99–100.

28   Provincial Oil Sands Study Group Proposals, May 6, 1974, GA CPA Box 34 File 532.

29   "Joint Response by the Canadian Petroleum Association and the Independent Petroleum Association of Canada to the Oil Sands Study Group Proposals of May 6, 1974," May 24, 1974, pp. 3–4, GA CPA Box 34 File 532.

30   Ibid., p. 2.

31   Doern and Toner, *Politics of Energy*, p. 173.

32   Foster, *Blue-Eyed Sheiks*, pp. 47–48.

33   Ibid.

34   Letter, Thomas P. Clarke, Canadian Petroleum Association Board of Directors to CPA Membership, January 11, 1972, GA CPA Box 13, File 188.

35   Ibid. CPA Letter to William Dickie, January 23, 1973, GA CPA Box 13, File 188.

36   Science Council Report, "Decision Making in the North: Oil Sands Case Study, November 1974," (Vancouver: Canadian Resourcecon Limited, 1974), p. 22.

37   Walwyn, Stodgell & Company made similar estimates in June 1973 but they concluded that processing would cost between six and seven thousand dollars per barrel per day. Walwyn, Stodgell & Co., Ltd., "A Look at World Energy: The Athabasca Tar Sands," June 1973, p. 28.

38   National Energy Board, "Potential Limitations of Canadian Petroleum Supplies," December 1972, GA CPA Box 44 File 534.

39 *Alberta Hansard, 1972*, March 29, 1972, vol. 18, p. 34.

40 Information in this table gathered from National Energy Board Reports for stated year.

41 Mandy J. Wahby, "Petroleum Taxation and Efficiency: The Canadian System in Question," *The Journal of Energy and Development* 9 (1984), p. 119.

42 Ibid.

43 Information contained in these tables gathered from National Energy Board Reports for stated year.

44 Alan J. MacFayden and Watkins (unpublished draft), pp. 34, 41, 47, 129. Stephen Clarkson and Christina McCall, *Trudeau and Our Times: The Heroic Delusion*, vol. 1 of 2 vols. (Toronto: McClelland & Stewart, 1994), pp. 168–69; Brian Scarfe, pp. 2, 9.

45 Letter, John Poyen, President, Canadian Petroleum Association, to CPA Membership, September 10, 1973, GA CPA Box 34, File 534.

46 Ibid.

47 For two very different views on the Syncrude negotiations, compare the accounts in Larry Pratt and Peter Foster. Pratt maintains that the oil companies swindled the provincial government and obtained a "sweetheart" deal. Foster, on the other hand, contends that the province drove a hard bargain and forced the company to capitulate to many of its demands. Pratt, *The Tar Sands*, pp. 115–78; Foster, *Blue-Eyed Sheiks*, pp. 97–99.

48 Yergin, *The Prize*, pp. 607–8. As Yergin explained, "production cuts would be more effective than a ban on exports against a single country, because oil could always be moved around, as had been done in the 1956 and 1967 crises. The cutbacks would assure that the absolute level of available supplies went down."

49 Thompson and Randall, *Ambivalent Allies*, p. 257.

50 Granatstein and Bothwell, *Pirouette*, p. 85. The suggestion that the multinational oil companies would allocate additional production to the United States at the expense of Canada was unfounded. The multinationals appreciated that any action they took would be subject to intense scrutiny and therefore decided on a policy of "equal misery" and allocated the same percentage of cutbacks from total supplies. See Yergin, *The Prize*, pp. 621–25.

51 *Hansard* December 7, 1974, 1st Session, 29th Parliament, p. 8478.

52 Ibid.

53 Granatstein and Bothwell, *Pirouette*, p. 86. The Canadian Development Corporation (1971) was a government-designed holding corporation that would attempt to mobilize investment capital to help "buy Canada back." Thompson and Randall, *Ambivalent Allies*, p. 255.

54 *Hansard* December 7, 1974, 1st Session, 29th Parliament, p. 8479.

55   In August 1971, Great Britain attempted to convert three billion dollars into gold and the Nixon administration responded by imposing a surcharge on imports and suspended the convertibility of dollars into gold, effectively devaluing the U.S. dollar. The moves took the Canadian government by complete surprise and officials in Ottawa believed Canada's inclusion in the measures was a mistake. The belief that Canadian interests warranted special treatment in Washington guided Canada's foreign policy. Nixon's moves seemed to indicate that the era of the "special relationship" was at an end. In 1972, Secretary of State for External Affairs, Mitchell Sharp, called for a reduction in U.S. economic and cultural influence in Canada. Increased nationalism on both sides of the border, noted Sharp in "Canada–U.S. Relations, Options for the Future," made a new approach necessary. Canada, he argued, could follow one of three alternatives. It could continue with the *status quo*, pursue a deliberate policy of greater integration with the United States, or select the third option: "develop and strengthen the Canadian economy and other aspects of its national life and in the process reduce the present Canadian vulnerability" to the vicissitudes of U.S. trade policy. Trade diversification (increased ties with the Japan and the Pacific Rim as well as trade agreements with the European Economic Community), and a new domestic industrial strategy that emphasized specialization and Canadian ownership would achieve these goals.

56   Bothwell, *Canada and the United States*, p. 104.

57   Fossum, *Oil, the State, and Federalism*, p. 34. See also Ted Greenwood, "Canadian-American Trade in Energy Issues," *International Organization* 28 (1974), pp. 689–710.

58   [Online]: Canadian Intellectual Property Office, <http://patents1.ic.gc.ca/> [8 August 2003].

59   Canadian Resourcecon, "Decision Making in the North," p. 24.

60   The Canadian government reached special arrangements for the northern tier of states to ensure they would not go without oil.

61   Bothwell, *Canada and the United States*, p. 120.

62   Trudeau did not specify in his December speech *when* domestic crude prices would rise or how the government would calculate a fair value for oil sands producers. Canadian prices remained a fraction of world levels, rising from approximately 59 per cent of the world price in 1974 to 83 per cent of world price in 1978 before the Iranian Revolution formally ended Ottawa's decision to sever the link between the "world" and "Canadian" price.

63   *GCOS Annual Report, 1974*, p. 2.

64   Canadian Resourcecon, "Decision Making in the North," p. 65.

65   *Hansard*, 1st Session, 29th Parliament, vol. 2, p. 7239.

66   Doern and Toner, *Politics of Energy*, p. 90.

67 The corporate culture of Calgary did not take kindly to the presence of the new company. Residents dubbed the company's new headquarters in downtown Calgary "Red Square" in record time. Children on the playgrounds of the city's schools chanted that Petro-Canada was an acronym for "Pierre Elliot Trudeau Rips Off Canada" to any unfortunate child whose parents happened to work for the company.

68 Alberta Department of Federal and Intergovernmental Affairs, "The Alberta Oil Sands Story" (Edmonton: Government Printer, 1974.).

69 AOSTRA is a crown corporation funded by the province to promote the development and use of new technology for oil sands and heavy oil production. AOSTRA operates primarily through shared-cost projects with industry. AOSTRA makes any technological developments that result from this co-operation available to any other user at fair market value.

70 Canadian Petroleum Association, "Oil Sands Development: Risks, Economics, Outlook," February 1975.

71 Canadian Petroleum Association, "Report of Meeting Between Alberta Government Oil Sands Study Group and Industry," April 29, 1974, GA CPA Box 34, File 531.

72 Richard Nixon, *RN: The Memoirs of Richard Nixon* (New York: Simon & Schuster, 1978), p. 984.

73 Canadian Petroleum Association, "Report of Meeting Between Alberta Government Oil Sands Study Group and Industry," April 29, 1974.

74 Letter, Sam Stewart, Chair CPA Oil Sand and Heavy Oil Committee, to Maurice Carrigy, Government of Alberta, May 9, 1974, GA CPA Box 34, File 532.

75 Kunz, *Butter and Guns*, p. 240.

76 Clayton Jones, "Alberta, Canada's Energy Giant, Turns Sand into Black Gold," *The Christian Science Monitor*, November 20, 1980, p. 3.

77 Canadian Petroleum Association, "Submission of the Canadian Petroleum Association to the Government of Alberta: Oil Sands Policy and Incentives," July 2, 1974, p. 2, GA CPA Box 34, File 532. Emphasis in original.

78 Ibid., p. 8.

79 In December 1978, the Province of Ontario would sell its 5 per cent share to PanCanadian Petroleum for $160 million.

80 Doern and Toner, *Politics of Energy*, p. 137.

81 Foster, *Blue-Eyed Sheiks*, p. 98.

82 Doern and Toner, *Politics of Energy*, p. 138.

83 Pratt, *The Tar Sands*, p. 102; Alberta Federal and Intergovernmental Affairs, *The Alberta Oil Sands Story*, p. 25.

84 Alberta Oil Sands Environmental Research Program, *Interim Report to 1978* (Edmonton: Government Printer, 1979), p. 2.

85   Alberta Oil Sands Environmental Research Program, *First Annual Report, 1975* (Edmonton: Government Printer, 1976), p. 4.

86   AOSERP, *Interim Report to 1978*, p. iv.

87   Honourable W.J. Yurko, "Address to the Engineering Institute of Canada Conference," Edmonton, Alberta, April 17, 1974.

88   Letter A.M. McIntosh to Honourable Don Getty, Minister of Energy Mines and Resources, February 24, 1977, GA CPA Box 59 File 925.

89   Canadian Petroleum Association, "An Assessment of Royalty Treatment and Other Factors Impacting Oil Sands Development," GA CPA Box 59 File 925.

90   Canadian Petroleum Association, "Confidential Notes on Meeting with Alberta/Federal Committee on 3rd Oil Sands Plant," May 18, 1977, GA CPA Box 59 File 925.

91   Canadian Petroleum Association, "1977 Annual Report of Activities Canadian Petroleum Association Oil Sands and Heavy Oil Committee," GA CPA Box 85 File 1022.

92   *Alberta Hansard*, October 26, 1978, p. 1539.

93   AOSERP, *Interim Report to 1978*, p. 2.

94   See, for example, Canadian Petroleum Association's Government Relations Contact Report of May 18, 1977, GA CPA Box 59 File 925.

Chapter 7

1   Doern and Toner, *Politics of Energy*, p. 6.

2   Peter Foster, *Sorcerer's Apprentices* (Toronto: HarperCollins, 1982), p. 19.

3   Gulf did not initiate its own oil sands project. Instead, it joined in several partnership agreements, including a share in the Syncrude project and part of the Alsands project.

4   See Chapter 5. Shell spent a great deal of time and effort in the 1960s researching and developing an *in situ* scheme that was ultimately turned down by the OGCB in 1963. The provincial Oil Sands Development Policy, and GCOS's preferential treatment thereunder, forced the company to shelve its oil sands plans indefinitely.

5   J.M.A Robinson and J.Y. Jessup, *The Alsands Project* (Calgary: Canadian Major Projects Association, 1988), pp. 14–16.

6   Material in this table is taken from John E. Feick, "Prospects for the Development of Minable Oil Sands," *Canadian Public Policy* IX (1983), p. 299; A. Janisch, "Oil Sands and Heavy Oil: Can they Ease the Energy Shortage?" *The United Nations Institute for Training and Research: The Future of Heavy Crude Oils and Tar Sands*, First International Conference, June 4–12, 1979, p. 40.

7   F.H. Allen, "The Canadian Oil Sands: A Race Against the Clock," *The United Nations Institute for Training and Research: The Future of Heavy Crude Oils and Tar Sands*. First International Conference, June 4–12, 1979, p. 31.

8   Foster, *Blue-Eyed Sheiks* (Toronto: Totem Books, 1979), p. 81. Technically, the Cold Lake deposit was classified as a heavy oil project by the ERCB because of the heavy oil's higher gravity. Generally speaking, the lower the API rating, the more asphalt the oil contains. While both are classified as "Heavy Crude Oils" by the American Petroleum Institute, oil sands bitumen typically measure below 10° API, while heavy oil measures 10° API or higher. Petroleum Communication Foundation. Available [Online]: <http://www.pcf.ab.ca/quick_answers/crude_oil/whatis.a> [12 April 2002].

9   Esso Resources Canada Limited, "Final Environmental Impact Assessment," Vol. 1, *Project Description*, 1979, p. 5.

10   Alan Michael Dooley, "The Cold Lake Oil Sands Project: An Economic Analysis," Master's Thesis, University of Alberta, 1981.

11   A. Janisch, "Oil Sands and Heavy Oil: Can They Ease the Energy Shortage?" p. 34.

12   Canadian Petroleum Association, Minutes of Oil Sands and Heavy Oils Committee Meeting, January 9, 1980, GA Box 85 File 1309; Allen, "The Canadian Oil Sands: A Race Against the Clock," p. 31.

13   Allen, "The Canadian Oil Sands: A Race Against the Clock," p. 32.

14   Iran's daily production was approximately 6 million barrels per day but dropped to 1.2 million in December 1978. The loss of Iranian production was offset by increase by other suppliers, resulting in a net loss of 2 million barrels per day.

15   Memo, Eliot Cutler, September 6, 1979, Carter Library, CF O/A 648, Box 45.

16   *Chemical Week*, June 6, 1979, p. 20

17   Ibid.

18   Jimmy Carter, *Keeping Faith* (New York: Bantam Books, 1982), p. 121.

19   Yergin, *The Prize*, p. 685.

20   Mohammed E. Ahrari, "A Paradigm of 'Crisis' Decision Making: The Case of Synfuels Policy," *British Journal of Political Science* 17 (1985), p. 82.

21   Yergin, *The Prize*, p. 689.

22   Arthur M. Okun, Treasury Department Report, May 24, 1979, Carter Library WHCF, O/A 540 Box 324.

23   Ibid.

24   Peter Foster, *Self-Serve*, p. 129.

25   In the spring of 1979, Trudeau's Liberal Party lost the federal election to Joe Clark's Progressive Conservative Party. Unfortunately, for Clark,

his party held a razor-thin majority in the House of Commons. The Conservatives held 136 seats to 114 for the Liberals, 26 for the New Democratic Party, and 6 for the Créditistes, making it essential that every Conservative Member of Parliament be present to vote in favour of government measures as well as garnering the support of a few MP's from the other parties. Clark and his Finance Minister, John Crosbie, unwittingly caused the demise of their government when they brought forth their first budget to a vote on the floor of the House of Commons without enough members present. The Clark government then lost a vote of non-confidence, and in the spring of 1980, Trudeau emerged from retirement to lead the Liberals to a resounding victory at the polls. "Winning politics depends not so much on how smart you are as on how dumb your opponent is," wrote Dave McIntosh. "Lucky Pierre Trudeau had Robert Stanfield and Joe Clark for antagonists." Dave McIntosh in Jack McLeod, ed., *The Oxford Book of Canadian Political Anecdotes* (Toronto: Oxford University Press, 1988), p. 194.

26   Doern and Toner, *Politics of Energy*, p. 31.

27   The Premiers of Quebec and Alberta, René Lévesque and Peter Lougheed, respectively, led the "Gang of Eight" in federal-provincial negotiations over the separation of powers and were, arguably, the two most powerful premiers in Canada. Lévesque's power derived from his ability to frustrate Trudeau's constitutional drive and from threatening Quebec separatism. Meanwhile, Alberta's tremendous wealth from natural resource revenues provided the province with an ample treasury to fund projects in other regions of the country, usurping the federal government's traditional role as the chief facilitator of income redistribution via equalization payments. Trudeau moved first against the Québécois separatists and mobilized the power of the federal government to win Quebec's 1980 sovereignty-association referendum for the "No" side with 60% of the vote. After suffering such an unequivocal defeat, the power and importance of the Parti Québécois of Réné Lévesque declined markedly. By moving against the source of Alberta's wealth, the oil industry, Trudeau hoped to remove the major source of Lougheed's independence.

28   Pierre Elliot Trudeau, *Memoirs* (Toronto: McClelland & Stewart, 1993), pp. 240–47.

29   Clarkson and McCall, *Trudeau and Our Times*, p. 160.

30   Trudeau, *Memoirs*, p. 272; Bliss, *Right Honourable Men*, p. 263.

31   Martin, *Chrétien: The Will to Win* (Toronto: Lester Publishing, 1995), p. 276.

32   Ron Graham, *One-Eyed Kings: Promise and Illusion in Canadian Politics* (Toronto: Totem Books, 1986), p. 58

33   Ibid., pp. 59–60.

34　Foster, *Blue-Eyed Sheiks*, p. 104.

35　Tom Kennedy, "Ottawa Blurs Hopes for Big Help from Oil Sands," *The Christian Science Monitor*, April 17, 1980, p. 11; *Chemical Week*, May 28, 1980, p. 16.

36　Trudeau, *Memoirs*, pp. 286–87. Readers should be wary of Trudeau's claims about the distribution of oil and gas revenues. A separate study by the Fraser Institute in 1981 argued that the federal government received 56.7 per cent, the industry 19.6 per cent, and the Province of Alberta 23.7 per cent. The study pointed out that Canadian consumers paid less than the world price for oil from 1973 onward, amounting to "a tax collected by Ottawa and distributed to Canadian consumers." Michael Walker, *Reaction: The National Energy Program* (Vancouver: The Fraser Institute, 1981), pp. xv-xvi.

37　Brian L. Scarfe, "Canadian Energy Prospects: Natural Gas, Tar Sands and Oil Policy," *Contemporary Policy Issues* III (1985), p. 2.

38　Many J. Wahby, "Petroleum Taxation and Efficiency: The Canadian System in Question," *The Journal of Energy and Development*, 9 (1), p. 111n2.

39　Canadian Petroleum Association, "The Selling of the National Energy Program and a Marketing Approach for the Alternative Oil and Gas Policy," October 2/3, 1983, GA CPA Box 106 File 1623.

40　"PM Rejected Offer of Final Talks on Oil," *The Calgary Herald*, October 28, 1980, p. A1.

41　Foster, *Sorcerer's Apprentices*, p. 74; Clarkson and McCall, *Heroic Delusion*, p. 173.

42　*EnFin*, when pronounced in French means "finally," and reflected the bureaucrats' attitudes toward the energy problem – "finally the energy problem would be solved!"

43　In mock tribute to the system's creator, Michael Pitfield.

44　In an effort to control government spending, Trudeau radically altered the government's budgetary system in the late 1970s with the creation of the Policy and Expenditure Management System. In its most basic form, the government instituted a series of policy "envelopes" that allotted a fixed annual sum to broad expenditure areas rather than to a specific department. Thus, External Affairs was lumped together with National Defense, and social development covered aspects of several different departments. So, each envelope was administered by a cabinet committee responsible for divvying up the contents of its envelope amongst themselves. If the ministers wanted to bring in new policies that cost more than their envelope contained, they could not go to the Finance Minister for more money. Instead, they would have to find it amongst themselves by agreeing among themselves to cut back on existing programs. Therefore, by giving *EnFin* its own envelope, Trudeau and Lalonde sharply limited the number of ministers privy to the government's energy strategy. Clarkson and McCall, *Heroic Delusion*, p. 173.

45 "Market Summary: TSE Oil and Gas Index," *The Calgary Herald*, October 22, 1980, p. C16; "Market Summary: TSE Oil and Gas Index," *The Calgary Herald*, October 28, 1980, p. C18.

46 "Alberta Economy Slowed by Unease Over Budget Plans," *The Calgary Herald*, October 28, 1980.

47 David Hatter, "First Shot Tonight in Energy 'War,'" *The Calgary Herald*, October 28, 1980, p. A1.

48 "'We Knew We Were Going to Get Hit.' Canada's National Energy Program Impact on Independent Petroleum Association of Canada in the 1980s; IPAC at 30." *Oilweek*, January 29, 1990, p. 18.

49 *Hansard, 1980*, vol. 4, p. 4186.

50 Doern and Toner, *Politics of Energy*, p. 2; G.C. Watkins and M.A. Walker, eds., *Reaction: The National Energy Program* (Vancouver: The Fraser Institute, 1981), p. xi.

51 The government's figures regarding American ownership and investment in the Canadian oil patch were controversial at the time, and remain so today. Michael Walker of the Fraser Institute assessed the level of U.S. ownership in Canada's petroleum sector to be approximately 40 per cent. As well, data from Statistics Canada tend to support assertions that levels of foreign ownership declined precipitously in the 1970s. Michael Walker, "The National Energy Program: An Overview of its Impact and Objectives," in Watkins and Walker, eds., *Reaction*, p. 27.

52 *The National Energy Program* (Ottawa: Minister of Supply and Services, 1980), p. 2.

53 Ibid., p. 7.

54 Ibid., pp. 24–25.

55 Ibid., p. 29.

56 *National Energy Board Annual Report, 1980*, p. 19.

57 *National Energy Program*, p. 39. The depletion allowance enabled producers to claim a federal income tax deduction equal to approximately one-third of all exploration, development, and certain capital expenditures related to oil and gas exploration. The depletion allowance for oil sands producers remained intact at 33.3 per cent.

58 Ibid., pp. 39–41.

59 Ibid., pp. 16–22.

60 Robert Lewis, "Lougheed Draws His Wagons into a Circle," *Maclean's*, November 10, 1980, p. 29; *House of Commons Debates*, IV (1980), p. 4196.

61 Editorial, "The National Energy Program," *The Calgary Herald*, October 29, 1980, p. A20.

62   Charles Lynch, "Flim Flam Man Envelopes Us In a Bank of Fog," *The Calgary Herald*, October 29, 1980.

63   The TSE Oil and Gas Index remained depressed throughout the winter and spring of 1980–81, typically a peak time for oil stocks because of increased drilling activity and consumption. The Oil and Gas Index finally returned to pre-budget levels in June, 1981 when it inched above 4971.89 points – just in time for the collapse of world crude oil prices that summer.

64   David Triguelro, "Petroleum, Gas Stocks Nosedive in Canada," *The Calgary Herald*, October 30, 1980; Gillian MacKay, "Double, Double, Oil and Trouble," *Maclean's*, November 10, 1980, p. 35.

65   Shane McCune, "Exploration Will Suffer: Oil Officials," *The Calgary Herald*, October 29, 1980, C1.

66   David Hatter, "Energy Executives Count Their Loses [*sic*]," *The Calgary Herald*, October 29, 1980.

67   *National Energy Program*, pp. 94–95. Emphasis mine.

68   "Energy Plan Slap In Face: Suncor Official," *The Calgary Herald*, October 20, 1980, p. D1.

69   Dave Pommer, "Federal Budget May Have Forced Lougheed Into Corner," *The Calgary Herald*, October 29, 1980, p. C2; Bob Bragg, "Oil Cut May be Lougheed's Top Weapon," *The Calgary Herald*, October 30, 1980, p. A1.

70   Peter Lougheed, *Transcript of Televised Address to Albertans, October 30, 1980*.

71   In any case, the Alsands Project Group already announced that it would not proceed without full assurances of world prices. Peter Eglington and Maris Uffelmann, "An Economic Analysis of Oilsands Policy in Canada – The Case of Alsands and Wolf Lake," Economic Council of Canada Paper #259 (Ottawa: Economic Council of Canada, 1984), p. 8.

72   Peter Lougheed, *Transcript of Televised Address to Albertans, October 30, 1980*.

73   Geoff White, "Door Open For New Oil Talks," *The Calgary Herald*, October 31, 1980; Trudeau, pp. 292–93.

74   *The Calgary Herald*, October 31, 1981.

75   "Budget and Federal Energy Policy Climax Exciting Year for Petroleum," *Oilweek*, December 1, 1980, p. 18.

76   Helliwell, p. 30.

77   "Can Canada Afford Its Oil Reform?" *Business Week*, November 17, 1980, p. 162.

78   Christopher Byron, "Canada's Barrel of Troubles; Inflation Leaps, the Dollar Sags and U.S. Oilmen Scream," *Time*, August 24, 1981, p. 55.

79   Robinson and Jessup, *Alsands Project*, p. 22.

80 Since some of the consortium's members qualified as "Canadian owned" or "Canadian controlled" they were subject to different levels of taxation by the federal government than American firms.

81 Doern and Toner, *Politics of Energy*, pp. 272–73.

82 Foster, *Sorcerer's Apprentices*, p. 179.

83 Eglington and Uffelmann, "Economic Analysis of Oilsands Policy," p. iv.

84 Foster, *Sorcerer's Apprentices*, p. 200.

85 Ibid., p. 198.

86 Canadian Petroleum Association, "Future Oil Sands Development in Alberta," December 1982, GA CPA Box 104 File 1610.

87 Yergin, *The Prize*, pp. 711–12.

88 Ibid. p. 714.

89 Ibid.

90 Foster, *Sorcerer's Apprentices*, p. 192.

91 The "Final Offer" presented by the province and the federal government. Eglington and Uffelmann, "An Economic Analysis of Oilsands Policy," pp. 16–17. Foster, *The Sorcerer's Apprentices*, pp. 200–201.

92 Ibid.

93 David G. Wood, *The Lougheed Legacy* (Toronto: Key Porter Books, 1985), p. 123.

94 G.W. Brandie, R.H. Clark, and J.C. Wiginton, "The Economic Enigma of the Tar Sands," *Canadian Public Policy* VIII (1982), p. 163.

95 Ibid.

96 Dr. Melon Speech, August 25, 1982, GA CPA Box 98 File 150.

97 Canadian Petroleum Association, "Future Oil Sand Development in Alberta," December 1982, p. 1, GA CPA Box 98 File 150.

98 Peter Foster, *Self-Serve*, pp. 224–25.

99 National Energy Board, *Canadian Energy: Supply and Demand, 1980–2000*, pp. 138–39.

100 National Oil Sands Task Force, "Canada's Oil Sands Industry: Yesterday, Today and Tomorrow" (Edmonton: Government Printer, 1995), pp. 8–9.

101 Yergin, *The Prize*, pp. 749–50.

102 "Canada's Energy Program Seen Hurting Energy Progress," *Oil and Gas Journal*, May 10, 1982, p. 94; "Canadian Energy self-sufficiency Doomed Until NEP is Abandoned," *Oil and Gas Journal*, June 21, 1982, p. 73; "Canadian Oil Industry Profits Down 34% in 1981," *Oil and Gas Journal*, August 30, 1982, p. 52; Eileen Doughty, "Canada's National Energy Program: How is it Affecting U.S. Firms?" *Business America*, January 24, 1983, p. 31.

103 Doern and Toner, *Politics of Energy*, p. 433.

104 Figure in 1999 dollars. Lorne Gunter, "Liberals Never Apologized for Sapping the West with NEP," *The Calgary Herald*, June 23, 1999, p. A19.

105 Pre-NEP projections envisioned that the sands would account for 20 per cent of Canada's annual crude production by 1990. The oil sands would only reach this level in 1995.

106 Yergin, "Conservation: The Key Energy Source," p. 139.

107 Ahrari, "A paradigm of 'Crisis' Decision Making," pp. 83–84.

Chapter 8

1 National Oil Sands Task Force, "The Oil Sands, A New Energy Vision For Canada" (Edmonton: Government Printer, 1995), p. 29.

2 Don Getty was among the first six Conservatives elected with Peter Lougheed to the provincial legislature in 1967. Getty served as Intergovernmental Affairs Minister in Lougheed's first cabinet and became Energy Minister in 1975. After a brief absence from politics, Getty became leader of the Provincial Progressive Conservative Party and premier on November 1, 1985.

3 Brooke Jeffrey, *Hard Right Turn* (Toronto: HarperCollins, 1999), p. 62.

4 Andrew Nikiforuk, "The New Quarterback," in Andrew Nikiforuk, Sheila Pratt, and Don Wanagas, eds., *Running on Empty: Alberta After the Boom* (Edmonton: NeWest Printing, 1987), pp. 118–19.

5 The first was the collapse of the Principal Group in 1987, a trust and financial institution. A subsequent investigation said one of the causes for the failure was "the wilful refusal of the [provincial] regulators to act effectively." The second failure was NovAtel, an early attempt by the province to create a high-tech communications corporation. NovAtel was a joint venture between the Nova Corporation and Alberta Government Telephones (AGT). The Nova Corporation began its life as Alberta Gas Trunk Line in 1954 to build, own, and operate Alberta's natural gas gathering and transmission facilities. Bob Blair, the son of Sydney M. Blair, joined the firm as executive vice-president in 1969 and became president a few years later. In 1988, Nova decided to sell its stake in NovAtel and AGT bought them out for $40 million. The company continued to operate until 1992 when the province decided to privatize the phone company. That is when citizens learned that NovAtel cost taxpayers approximately $566 million in guaranteed loans.

6 Frank Dabbs, *Ralph Klein: A Maverick Life* (Vancouver: Greystone Books, 1995), p. 75.

7 A lawyer and former CEO of Algoma Steel, Mulroney first ran for the leadership of the Progressive Conservative Party in 1976 but lost to Joe Clark. After Clark's ignominious defeat at the hands of the Liberals,

the Conservatives looked for a new leader in 1983. A skilful manager, Mulroney focused on healing party wounds; he defeated Clark on the final ballot at the nomination convention.

8   Clarkson, *Uncle Sam and Us*, p. 28.

9   Peter C. Newman, *Canadian Revolution*, p. 256.

10   Granatstein and Hillmer, *For Better or For Worse*, p. 278.

11   Ibid., p. 301.

12   Ibid., p. 296.

13   Hillmer and Granatstein, *Empire to Umpire*, pp. 335–36; Donald S. Macdonald, "Leap of Faith," in Ian MacDonald, *Free Trade: Risks and Rewards* (Montreal: McGill-Queen's University Press, 2000), p. 52.

14   Bliss, *Right Honourable Men*, p. 295.

15   Gerald Caplan, Michael Kirby, and Hugh Segal, *Election: The Issues, the Strategies, the Aftermath* (Scarborough: Prentice-Hall, 1989); John Duffy, *Fights of Our Lives*, pp. 311–57.

16   Michel Duquette, "Domestic and International Factors Affecting Energy Trade," in Stephen J. Randall and Herman Konrad, eds., *NAFTA in Transition* (Calgary: University of Calgary Press, 1995), p. 303.

17   Thompson and Randall, *Ambivalent Allies*, p. 289.

18   Duquette, "Factors Affecting Energy Trade," pp. 303–4.

19   Caplan, Kirby, Segal, *Election*, p. 37.

20   Hyman Soloman, "Politics Oils Gears of Mega-Projects," *Financial Post*, January 18, 1988, p. 15.

21   Mansell and Percy, *Strength in Adversity*, pp. 9, 11.

22   Canadian Press, "Progress Made Toward Energy Agreement, Ministers Say," *The Montreal Gazette*, March 26, 1985, p. C9.

23   Canadian Press, "18,000 May Lose Jobs Because of Slump, Say Oil Producers," *The Ottawa Citizen*, March 29, 1986, p. D21.

24   Canadian Press, "Imperial Oil Earnings Hit the Skids," *The Ottawa Citizen*, July 23, 1986, p. E1; David Climie, "Collapsing Oil Prices Fall Hard on Stocks," *The Ottawa Citizen*, June 25, 1986, p. C11.

25   Canadian Press, "Oil Price Fall Hits Alberta the Hardest," *The Ottawa Citizen*, August 2, 1986, p. E7.

26   Alexander McEachern, *Alberta Hansard*, Vol. 1., April 14, 1988, p. 457.

27   Quoted by Sheldon Chumir, *Alberta Hansard*, Vol. 1, March 24, 1987, p. 325.

28   Dr. Neil Webber, *Alberta Hansard*, March 24, 1987, pp. 318–19.

29   Energy Minister Rick Orman, *Alberta Hansard*, April 25, 1991. Available [Online]: <http://isys.assembly.ab.ca:8080/ISYSquery/IRL57B1.tmp/4/doc> [13 August 2003].

30   Neil Webber, *Alberta Hansard*, vol. 2, May 31, 1988, p. 1375.

31   Terence Corcoran, "Dangerous Fusion of Politics and Business," *Financial Post*, September 28, 1988, p. 13.

32   Thomas Kierans, et al, *Energy and Canadians Into the 21st Century* (Ottawa: Government Printer, 1988), p. A11.

33   Ibid., p. 6.

34   Ibid., p. 47.

35   William J. Baumol and Edward N. Wolff, "Subsidies to New Energy Sources: Do They Add to Energy Stocks?" *Journal of Political Economy*, 89 (1981), pp. 891–913.

36   David Hatter, "Masse Gives Report the Cold Shoulder," *Financial Post*, August 5, 1988, p. 6.

37   Lynn Hunter, *Hansard*, vol. 2, June 8, 1989, pp. 2759–2760.

38   Frank Dabbs, "OSLO Process Points to Age of Hydrogen Energy," *Financial Post*, January 18, 1990, p. 15.

39   Dennis Hryciuk, "Doubts Linger About Future of OSLO Project; Major Partner Losing Interest," *The Edmonton Journal*, December 5, 1989, p. D10; Dennis Hryciuk, "Oil Prices Jeopardize OSLO Project," *The Edmonton Journal*, December 2, 1989, p. G1.

40   Fossum, *Oil, the State and Federalism*, p. 203.

41   Don Getty, *Alberta Hansard*, vol. 3, August 3, 1989, p. 1234.

42   Michael Wilson, *Hansard*, vol. 6, February 20, 1989, p. 8599.

43   Peter Morton, "Austerity Move Puts Petro-Canada On the Block," *The Oil Daily*, February 22, 1990, p. 1; Frank Dabbs, "Project Partners Assess Options," *Oilweek*, March 5, 1990, p. 5.

44   Tamsin Carlisle, "Where Petro-Can Wants to Go," *Financial Post*, February 26, 1990, p. 19; Foster, *Self Serve*, p. 274.

45   Lynda Shorten, "Getty Warns of Energy Crisis if OSLO Dies," *The Edmonton Journal*, February 22, 1990, p. A1; Editorial, "Low-Key Cries of Pain," *The Edmonton Journal*, February 22, 1990, p. A14.

46   Frank Dabbs, "Alberta's OSLO Response Threatens Petro-Canada," *Financial Post*, February 23, 1990, p. 3.

47   Frank Dabbs, "Ontario will Examine OSLO Deal," *Oilweek*, April 9, 1990, p. 5.

48   Rick Pedersen, "OSLO Chief Admits Project Timing Poor," *The Edmonton Journal*, April 18, 1991, p. C5.

49   Dennis Hryciuk, "Canada Imports Oil for the First Time in 8 Years," *The Edmonton Journal*, June 5, 1990, p. D7; Canadian Press, "Foreign Oil Flooding into Ontario," *The Edmonton Journal*, March 3, 1990, p. D3; Frank Dabbs, "Are Canadian Producers Truly Competitive?" *Oilweek*, January 8, 1990, p. 3.

50   Dennis Hryciuk, "Syncrude Oil Sands Pullouts View[ed] As A Bad Omen for OSLO's Future," *The Edmonton Journal*, May 15, 1990, p. B4.

51   Frank Dabbs, "Oil Sands the Key to Energy Security," *Oilweek*, June 18, 1990, p. S2; Rick Pederson, "400 Jobs to be Trimmed at Syncrude;

Administration the Target As Firm Seeks $15-Per-Barrel Cost of Production," *The Edmonton Journal*, July 9, 1991, p. D9.

52   Tamsin Carlisle, "Oil Sands Development Slows: The Flight From Heavy Oil," *Financial Post*, May 14, 1990, p. 1.

53   Marc Lisac, *The Klein Revolution* (Edmonton: NeWest Printing, 1995) p. 51.

54   Ibid., p. 64.

55   Peter Morton, "$4 Billion Canadian OSLO Oil Sands Project Dead," *The Oil Daily*, November 3, 1992, p. 2.

56   *Alberta Hansard*, April 4, 1991. Available [Online]: <http://isys.assembly.ab.ca:8080/ISYSquery/IRL57B1.tmp/1/doc> [13 August 2003].

57   Richard Helm, "Gulf Crisis Called Stimulus for OSLO; 'Better' Chance for Oil Sands Project," *The Edmonton Journal*, August 25, 1990, p. A8; Kristin Goff, "Alberta Oil Patch Starting to Strut Again," *The Ottawa Citizen*, August 25, 1990, p. F1; Jack Danylchuk, "The Windfalls of War; Trouble in the Gulf Means Profit in the Oilfields," *The Ottawa Citizen*, p. I8; Jack Danylchuk, "Kuwait's Loss Alberta's Gain: Invasion Means Bonanza for Oil Patch, Could Bring Extra $2 Billion to Government," *The Edmonton Journal*, August 11, 1990, p. G1; Tamsin Carlisle, "Canada's Ailing Oil Patch Could Reap Billions in Spoils War," *Financial Post*, August 3, 1990, p. 4.

58   A former teacher and television reporter, Ralph Klein entered municipal politics in 1980, serving three terms as Calgary's mayor. His down-to-earth style – citizens knew they could speak with Klein one-on-one if they visited the bar in the King Edward Hotel – appealed to Calgarians. Klein stepped down as mayor in 1989 to run as a Conservative candidate in that year's provincial election. After his successful campaign, Premier Getty appointed Klein Environment Minister, launching Klein's career in provincial politics.

59   Dabbs, *Ralph Klein*, p. 107.

60   Ibid., p. 116.

61   Among Klein's biographers, there is some dispute about whether or not the Klein government possessed an overarching ideological blueprint to guide actions. Frank Dabbs wrote in his 1995 biography that Klein and his chief lieutenants were heavily influenced by two books in particular: David Osbourne and Ted Gaebler's *Reinventing Government* (the authors are public sector management consultants) and former New Zealand Finance Minister Sir Roger Douglas's *Unfinished Business*. Conversely, Canadian political scientist, former senior bureaucrat, and federal Liberal candidate Brooke Jeffrey argues that Klein acted without the benefit of an overall plan. Instead, she maintains that Klein's government tended "to act first and think later, impulsively adopting elements

of the neo-conservative agenda without having an overall strategy." Don Martin's 2002 biography tends to reinforce Jeffrey's argument. While Martin agrees that the books mentioned by Dabbs were closely read by elements within Klein's government, the premier never read such books himself. Jeffrey, *Hard Right Turn*, p. 126; Dabbs, p. 110; Don Martin, *King Ralph* (Toronto: Key Porter Books, 2002), p. 140.

62 Lisac, *The Klein Revolution*, p. 43. Lisac never fully made the connection between the "Klein Revolution" and globalization. Instead, he drew analogies between fascist Italy and Alberta.

63 Daniel Yergin and Joseph Stanislaw, *The Commanding Heights: The Battle Between Government and the Marketplace that Is Remaking the Modern World* (New York: Simon & Schuster, 1998), p. 105.

64 Ibid.

65 Gil Troy, "Ronald Reagan" in Alan Brinkley and Davis Dyer, eds., *The Reader's Companion to the American Presidency* (Boston: Houghton Mifflin, 2000), p. 493.

66 Ibid.

67 Thomas Friedman, *The Lexus and the Olive Tree* (New York: Anchor Books, 2000), p. 104.

68 Although Brian Mulroney did acknowledge the need for reduced taxes, balanced budgets, and free trade, he only succeeded in implementing the last of these items during his nine years in office.

69 Bliss, *Right Honourable Men*, p. 270.

70 Maryse Robert, *Negotiating NAFTA: Explaining the Outcome in Culture, Textiles, Autos, and Pharmaceuticals* (Toronto: University of Toronto Press, 2000), p. 31.

71 Thompson and Randall, *Ambivalent Allies*, p. 292.

72 The political fortunes of the Progressive Conservative Party received a shocking blow in the early 1990s. After a series of stunning rebukes by voters regarding Mulroney's changes to the constitution, Mulroney stepped down as leader in 1993 and turned control of the Conservatives to Kim Campbell. Fed up with the perceived arrogance of Brian Mulroney and the economic downturn, the Conservative coalition disintegrated and resulted in the election of only two Conservatives to the House of Commons. In his place, Canadians overwhelmingly turned to the Liberal Party headed by former Trudeau Cabinet Minister Jean Chrétien.

73 Friedman, *The Lexus and the Olive Tree*, p. 106.

74 National Oil Sands Task Force, "The Oil Sands: A New Energy Vision for Canada," p. 6.

75 National Oil Sands Task Force, "Securing a Sustainable Future for Canada's Oil Sands Industry" (Edmonton: Government Printer, 1995), p. 1.

76 Alan Boras, "Teamwork Reigns in Oilsands," *The Calgary Herald*, May 21, 1995, p. F1.

77 National Oil Sands Task Force, "Securing a Sustainable Future for Canada's Oil Sands Industry," p. 23.

78 Ibid., p. 24.

79 Data taken from National Oil Sands Task Force, "A Science and Technology Strategy for Canada's Oil Sands Industry" (Edmonton: Government Printer, 1995), p. 5. Monetary figures given in constant 1995 dollars.

80 John Chadwick, "Mining Alberta's Oil Sands," *Mining Magazine*, August, 1998, p. 58.

81 National Energy Board, "Canada's Oil Sands: A Supply and Market Outlook to 2015" (Ottawa: Government Printer, 2000), pp. 22–23.

82 Ibid., pp. 27–28.

83 Ibid.

84 Alberta Oil Sands Technology and Research Authority: *AOSTRA: A 15-Year Portfolio of Achievement* (Edmonton: Government Printer, 1991), p. 11.

85 Oil Sands Task Force, "The Oil Sands: A New Vision For Canada," p. 24.

86 Ibid., p. 26.

87 Alberta Government News Release, November 30, 1995. [Online]: <http://www.energy.gov.ab.ca/com/Room/News+Releases/1995/NOV-30-1995.htm> [15 April 2002].

88 Robert Mitchell, Brad Anderson, Marty Kaga, and Stephen Eliot, "Alberta's Oil Sands: Update on the Generic Royalty Regime," Alberta Department of Energy. Available [Online]: <http://www.energy.gov.ab.ca/com/Sands/Royalty+Info/Royalty+Related+Info/Update+on+the+Generic+Royalty+Regime.htm> [15 April 2002].

89 Ibid.

90 Before this decision, *in situ* projects were not treated as mining projects, meaning that they did not qualify for a special tax deduction. The 1996 decision effectively levelled the playing field by placing *in situ* developers on equal footing with mining projects.

91 National Energy Board, "Canada's Oil Sands," p. 40.

92 Energy and Utilities Board, "Alberta's Energy Resources: 1997 in Review," (Edmonton: Government Printer, 1998), pp. 5–6.

93 Energy and Utilities Board, "Historical Overview," p. 9.

94 Ibid.

95 National Energy Board, "Canada's Oil Sands," p. 58.

96 Ibid., p. 78.

97 Ibid.

98 Peter Foster, "Mr. Klein Goes to Washington," *The National Post*, June 13, 2001. [Online]: <http://www.nationalpost.com/story.html?f=/stories/20010613/589621.html> [13 June 2001].

99 David Suzuki Foundation, "Meeting U.S. Energy Demands Worsens Climate Change." February 23, 2001. [Online]: http://www.davidsuzuki.org/campaigns_and_programs/climate_change/news_releases/newsclimatechange03010102.asp> [9 November 2001].

100 Observers and political commentators have long argued that Chrétien's commitment to Kyoto is the product of his attempt to appear as a greater liberal than former President Bill Clinton and to lay the basis for his legacy. CBC News. The National Online, October 27, 1997, [Online]: <http://tv.cbc.ca/national/trans/T971028.html> [2 October 2002]; Terence Corcoran, "The Long, Sad Legacy of Kyoto," *The National Post*, April 16, 2002, [Online]: <http://www.nationalpost.com/> [17 April 2002]; Jeffrey Simpson, "Why Canada's on the Kyoto Spot," *The Globe and Mail*, April 19, 2002, [Online]: <http://www.globeandmail.com/> [19 April 2002].

101 Prime Minister of Canada's Office, "Prime Minister Welcomes New Initiatives with the European Union," June 21, 2001. [Online]: <http://pm.gc.ca/ default.asp? Language=E&Page=newsroom&Sub=newsreleases&Doc=newinitiativeseu.20010621_e.htm> [22 September 2001].

102 Environment Canada, "Remarks of Environment Minister David Anderson, Montreal, PQ." March 29, 2001, [Online]: <http://ec.gc.ca/minister/speeches/2001/010328-2_s_e.htm> [12 November 2001]; Natural Resources Canada. "Government of Canada Commits to Greenhouse Gas Reductions." June 6, 2001, [Online]: <http://www.nrcan.gc.ca/css/imb/hqlib/200140e.htm> [12 November 2001].

103 *Hansard*, House of Commons, *Standing Committee on Aboriginal Affairs, Northern Development and Natural Resources*, May 17, 2001. Available [Online]: <http://www.parl.gc.ca/common/Chamber_House_Debates.asp?Language=E&Parl=37&Ses=1> [12 November 2001].

104 Ibid.

105 *Hansard*, House of Commons Debates, 37th Parliament, 1st Session. April 25, 2001. Available [Online]: <http://www.parl.gc.ca/common/Chamber_House_Debates.asp?Language=E&Parl=37&Ses=1> [1 November 2001].

106 *Hansard*, House of Commons Debates, 37th Parliament, 1st Session. March 31, 2001. Available [Online]: <http://www.parl.gc.ca/common/Chamber_House_Debates.asp?Language=E&Parl=37&Ses=1> [1 November 2001].

107 *Hansard*, House of Commons Debates, 37th Parliament, 1st Session. April 24, 2001. Available [Online]: <http://www.parl.gc.ca/common/

Chamber_House_Debates.asp?Language=E&Parl=37&Ses=1> [1
November 2001].

108   Ibid.

109   Ibid.

110   Ibid.

111   *Hansard,* House of Commons, *Standing Committee on Aboriginal Affairs,
Northern Development and Natural Resources,* May 17, 2001. [Online]:
<http://www.parl.gc.ca/infocomdoc/37/1aanr/meetings/evidence/
aanrev/6-e.htm> [8 November 2001].

112   In the early 1990s, the Canadian Petroleum Association and the
Independent Petroleum Association of Canada merged to form the
Canadian Association of Petroleum Producers.

113   Canadian Association of Petroleum Producers, "Oil and Natural Gas
Strategies for North American Energy Markets," April 2001. [Online]:
<http://www.capp.ca> [12 November 2001].

114   Abraham McLaughlin, "Fading Talk of Energy 'Crisis' Hurts Bush
Plan," *The Christian Science Monitor,* July 20, 2001. Available [Online]:
<http://www.csmonitor.com/durable/2001/07/20/p3s1.htm> [1 August
2001].

115   Peter Morton, "Klein and Cheney Talk Oil, Fishing Instead of
Politics," *The National Post,* June 15, 2001. [Online]: <http://
www.nationalpost.com/story.html?f=/stories/20010615/592378.html>
[16 June 2001].

116   Foster, "Mr. Klein Goes to Washington."

117   Frank Dabbs, "Senator Shows the Way for Oil Ministers," *The Calgary
Herald,* October 26, 2001, p. F2; author interview, Senator Nicholas
Taylor, November 6, 2001.

118   Ibid.

119   Ali Rodriguez Araque, "Geopolitics of International Oil Industry and
OPEC," Speech, Vienna, Austria, October 5, 2001, available [Online]:
<http://www.opec.org> [7 November 2001]; Adel Khalid Al-Sebeeh,
"Interdependence, No Longer a Choice, Now an Imperative," Speech,
23rd Oxford Energy Seminar, September 13, 2001, available [Online]:
<http://www.opec.org> [7 November 2001]; Ali Rodriguez Araque,
"OPEC and the Oil Market in the Near Future," Speech, Montreux
Energy Roundtable XII, May 28-30, 2001, available [Online]: <http:
//www.opec.org> [7 November 2001].

120   Senate of Canada, *Proceedings of the Standing Senate Committee on
Energy, the Environment, and Natural Resources. Issue 3 – April 24, 2001.*
[Online]:   <http://www.parl.gc.ca/37/1parlbus/commbus/senate/com-
e/03eva-e.asp> [8 November 2001].

121   American Petroleum Institute. "DOE Discusses Energy Policy in
Light of September 11 Events," October 18, 2001. [Online]: <http:

//api-ep.api.org/printerformat.cfm?   ContentID=E78E5769-7B52-4A8D-95F3C81CE472936F> [8 November 2001].

122   Neela Banerjee, "Fears, Again, of Oil Supplies at Risk," *The New York Times*, October 14, 2001. [Online]: <http://www.nytimes.com/2001/10/14/business/14OILL.html?pagewatned=print> [15 October 2001].

123   Neela Banerjee, "The High, Hidden Cost of Saudi Arabian Oil," *The New York Times*, October 21, 2001. [Online]: <http://www.nytimes.com/2001/10/21/weekinreview/21BANE.html?pagewanted=print>[22 October 2001].

124   David B. Ottaway and Robert G. Kaiser, "After Sept. 11, Severe Tests Loom for Relationship." *The Washington Post*, February 12, 2002. [Online]: <http://www.washingtonpost.com/ac2/wp-dyn?pagename=article&node=&contentId=A60502-2002Feb11> [12 February 2002].

125   Peter Verburg, "Land of the Giants," *Canadian Business*, April 15, 2002. Available [Online]: <http://www.lexis-nexis.com> [19 April 2002].

126   Alan Toulin, "Will Ratify Kyoto 'One Day' Chrétien Says," *The National Post*, April 16, 2002. [Online]: <http://www.nationalpost.com/> [17 April 2002].

127   Steven Chase, "EU to Fight Ottawa Bid for More Kyoto Credits," *The Globe and Mail*, April 12, 2002. [Online]: <http://www.globeandmail.com/servlet/GIS.Servlets.HTMLTemplate?tf=tgam/search/tgam/SearchFullStory.html&cf=tgam/search/tgam/SearchFullStory.cfg&configFileLoc=tgam/config&encoded_keywords=kyoto&option=&start_row=5&current_row=5&start_row_offset1=&num_rows=1&search_results_start=1> [13 April 2002].

128   Canadian Broadcasting Corporation, "Canadian business groups say Kyoto agreement will cost 2.5% of GDP," March 4, 2002. Available [Online]: <http://www.cbc.ca/stories/2002/03/04/kyoto_020304> [5 March 2002].

129   Canadian Manufacturer's and Exporters, "Pain Without Gain: Canada and the Kyoto Protocol," no date. [Online]: <http://www.cme-mec.ca/kyoto/documents/kyoto_release.pdf> [19 April 2002].

Chapter 9

1   Address by Prime Minister Jean Chrétien at the World Summit on Sustainable Development, September 2, 2002 Johannesburg, South Africa. [Online]: <http://pm.gc.ca/ default.asp? Language=E&Page=newsroom&Sub=speeches&Doc=sommetmondialpourledéveloppementdurable2.20020902_e.htm> [16 September 2002].

2   Kyoto's provisions specifically address three kinds of emissions - carbon dioxide (CO_2), nitrous oxide (NO), and methane (CH_4). Much of the public debate on Kyoto in Canada has focused on CO_2 exclusively.

3   Robert Priddle, "The Future Energy Supply of Society: Options, Risks, and Choices," March 13, 2000. [Online]: <http://www.iea.org/new/speeches/priddle/2000/stockh.pdf> [21 January 2003].

4   Ibid.; Faith Birol, "World Oil Outlook to 2030," November 25, 2002. [Online]: <http://www.iea.org/workshop/amos/birol.pdf> [21 January 2003].

5   Ross McKitrick and Randal M. Wigle, "The Kyoto Protocol: Canada's Risky Rush to Judgement," (Toronto: C.D. Howe Institute, 2002), p. 3. Available [Online]: <http://www.cdhowe.org/pdf/commentary_169.pdf> [1 October 2002].

6   Government of Alberta, "Global Greenhouse Gas Emissions, 2000." Available [Online]: <http://www.gov.ab.ca/home/kyoto/Display.cfm?id=4> [3 October 2002].

7   Meaning that Kyoto will account for only 30 per cent of worldwide CO2 emissions.

8   Steven Chase and Alanna Mitchell, "Chrétien Could Deliver Kyoto to World," *The Globe and Mail*, August 28, 2002. [Online]: <http://www.globeandmail.com/servlet/ArticleNews/PEstory/TGAM/20020828/UKYOTN/Headlines/headdex/headdexInternational_temp/5/5/24/> [28 August 2002].

9   Mexico ratified the Kyoto protocol on September 7, 2000 but its status as an Annex II country does not commit the country to emissions reductions.

10   Lisa Schmidt, "EU Said Unlikely to Meet Kyoto Targets," *The Calgary Herald*, October 3, 2002, p. D2.

11   Roy MacLaren, "Wanted: EU Trading Partners," *The Globe and Mail*, August 16, 2002. [Online]: <http://www.globeandmail.com/servlet/ArticleNews/PEstory/TGAM/20020816/COEU/Headlines/headdex/headdexComment_temp/1/1/6/> [3 October 2002].

12   John Vidal, Charlotte Denny, and Larry Elliott, "Secret Documents Reveal EU's Tough Stance on Global Trade," *The Guardian* (U.K.), April 17, 2002. Available [Online]: <http://www.guardian.co.uk/international/story/0,3604,685650,00.html> [4 October 2002].

13   *Hansard*, House of Commons, October 24, 2002. Available [Online]: <http://www.parl.gc.ca/37/2/parlbus/chambus/house/debates/014_2002-10-24/han014_1345-E.htm> [10 January 2003].

14   Scott Haggett, "Kyoto Threatens $4B Oil Project," *The Calgary Herald*, September 28, 2002, available [Online]: <http://www.canada.com/search/site/story.asp?id=7C52AEB8-A00E-4302-9D82-0927D2F5AAA2> [3 October 2002]; Bryant Avery, "Kyoto Costs put Projects in Question," *The Edmonton Journal*, January 14, 2003, [Online]: <http://www.canada.com/search/story.aspx?id=16bda0ba-c4ea-4d11-9d46-43fd73302039> [14 January 2003].

15 Mark Heckathorn, "CNR Delays Oilsand Project," *Natural Gas Week*, June 21, 2002. Available [Online]: <www.lexis-nexis.com> [24 June 2002].

16 Marlie Burtt, "Design of Fiscal Terms and Incentives to Encourage Investment," November 26, 2002. Available [Online]: <http://www.iea.org/workshop/amos/Burtt.pdf> [22 January 2003].

17 Ibid.

18 Bernard Bigras, "Why Would an Equitable Method of Sharing the Burden be Good For Europe and Bad for Canada?" *The Hill Times*, January 20, 2003. [Online]: <http://www.thehilltimes.ca/2003/january/20/pb_bigras.> [25 January 2003].

19 CBC News, "Automakers Exempted From Kyoto Requirements," January 3, 2003. Available [Online]: <http://cbc.ca/storyview/CBC/2003/01/03/kyotocars030103> [10 January 2003].

20 *Hansard*, House of Commons, October 3, 2002. Available [Online]: <http://www.parl.gc.ca/common/Chamber_House_Debates.asp?Language=E&Parl=37&Ses=1> [10 January 2003].

21 David Suzuki Foundation, "Kyoto Critical for Clean Air, Health, Say Physicians," September 25, 2002. Available [Online]: <http://www.davidsuzuki.org/Campaigns_and_Programs/Climate_Change/News_Releases/newsclimatechange09250201.asp > [26 September 2002]. In response to a question by Opposition Leader Steven Harper, Environment Minister David Anderson explained that CO2 reductions were necessary to improve air quality. "Pollutants emitted into the atmosphere, such as nitrous oxide and sulphur dioxide, need heat to make the smog that we see so often in Canada," *Hansard*, House of Commons, October 11, 2002, Available [Online]: <http://www.parl.gc.ca/common/Chamber_House_Debates.asp?Language=E&Parl=37&Ses=1> [14 January 2003]; Tom Spears, "Accord Won't Save Lives: Experts," *The Ottawa Citizen*, September 27, 2002, [Online]: <http://www.canada.com/search/site/story.asp?id=CA29A416-3144-475A-A197-841CD56E853B> [3 October 2002].

22 Bill Curry, "Canadians 'Ignorant' of Energy Facts," *The National Post*, September 16, 2002, [Online]: <http://www.nationalpost.com/search/site/story.asp?id=39787A0A-8ACC-4666-8E39-298354C51DB8> [26 September 2002]; Canadian Press, "Poll Shows Kyoto Support Drops with Awareness," September 6, 2002, available [Online]: <http://www.ctv.ca/servlet/ArticleNews/story/CTVNews/1031300074132_37//> [29 September 2002].

23 Bruce Cheadle, "Canada to Sign Kyoto, but Won't Abide By It," *The Toronto Star*, September 5, 2002. [Online]: <http://www.thestar.com/NASApp/cs/ContentServer?pagename=thestar/Layout/Article_Type1&c=Article&cid=1026144897361&call_page=TS_News&call_pageid=968332188492&call_pagepath=News/News&col=968793972154> [30 October 2002].

24   CBC News, "Kyoto Implementation Plan will be 10 Years in the Works: PM," September 25, 2002. Available [Online]: <http://cbc.ca/storyview/CBC/2002/09/24/kyoto_plan020924> [3 October 2002].

25   Patrick Brethour and Steven Chase, "Kyoto Cited as Oil Sands Project Put on Ice," *The Globe and Mail*, September 20, 2002. [Online]: <http://www.globeandmail.com/servlet/GIS.Servlets.HTMLTemplate?current_row=26&tf=tgam/search/tgam/SearchFullStory.html&cf=tgam/search/tgam/SearchFullStory.cfg&configFileLoc=tgam/config&encoded_keywords=kyoto&option=&start_row=26&start_row_offset1=0&num_rows=1&search_results_start=21&query=kyoto> [26 September 2002]. On January 14, 2003, Koch Industries shelved its U.S.$ 2.2 billion oil sands project indefinitely after failing to find a partner to limit Koch's exposure. [Online]: <http://story.news.yahoo.com/news?tmpl=story&u=/nm/20030114/wl_canada_nm/canada_energy_truenorth_col_1> [14 January 2003].

26   Sylvia LeRoy and Jillian Frank, "Kyoto and the Constitution," *Fraser Forum*, October 2002, pp. 5-6, [Online]: <http://www.fraserinstitute.ca/admin/books/chapterfiles/Kyoto%20and%20the%20Constitution-pp5-6.pdf>[3 January 2003]; Allan Gotlieb and Eli Lederman, "Ignoring Provinces is Not Canada's Way" *National Post*, January 3, 2003, [Online]: <http://www.nationalpost.com/search/site/story.asp?id=E55A20B1-96AC-4AEB-ABD9-8E3823A863A7> [3 January 2003].

27   Paul Haavardsrud, "Kyoto Deal Would Savage Oil, Gas Stocks: Analysts," *Financial Post*, September 5, 2002. [Online]: <http://www.nationalpost.com/search/site/story.asp?id=02D604C6-3688-4851-B917-841F9CCD0DEC> [17 September 2002].

28   Gwyn Morgan, "The Real Chill from Kyoto," *The National Post*, September 26, 2002. [Online]: <http://www.nationalpost.com/search/site/story.asp?id=0BBD9EE3-4AAB-418C-A9A8-E944AF4B7DBB> [26 September 2002].

29   Robert Benzie, "Offer Details or Ontario will Opt Out, Eves Warns Ottawa," *The National Post*, October 2, 2002, [Online]: <http://www.nationalpost.com/search/site/story.asp?id=ACA2DF38-A9ED-4EE2-AF58-DE38EF444205> [2 October 2002]; Angela Hall, "Opposition Wants Legislature Recalled to Debate Kyoto Plan," *The Regina Leader-Post*, October 3, 2002, [Online]: <http://www.canada.com/search/site/story.asp?id=55F908D5-6204-41B7-80AD-B9672687E605> [3 October 2002].

30   Canadian Press, "Time for Debate On Kyoto: Sask Party," October 2, 2002. [Online]: <http://www.canada.com/search/site/story.asp?id=ECAD965D-8B86-4378-85DA-548D0CB7D266> [3 October 2002].

31   House of Commons, Standing Committee on Finance, Evidence, November 6, 2002. [Online]: <http://www.parl.gc.ca/InfoComDoc/37/

2/FINA/Meetings/Evidence/FINAEV19-E.HTM#Int-323681 > [16 January 2003].

32   Ibid.

33   The remainder would be met from a few other resources, including production from the East Coast. National Energy Board, "Canada's Oil Sands: A Supply and Market Outlook to 2015" (Ottawa: Government Printer, 2000), pp. 46-48.

34   National Energy Board, "Crude Oil and Petroleum Products: Export Statistics." [Online]. <http://www.neb.gc.ca/stats/oil/index_e.htm> [22 January 2003].

35   National Oil Sands Task Force, *Environmental Report: Securing a Sustainable Future for Canada's Oil Sands Industry* (Edmonton: Alberta Chamber of Resources, 1995); Canadian Association of Petroleum Producers, "Climate Change Policy Position," November 2002, p. 2.

36   Steven Chase, "Ottawa to Limit Emission Reduction Burden," *The Globe and Mail*, December 18, 2002, [Online]. <http://www.globeandmail.com/servlet/ArticleNews/business/RTGAM/20021218/wkyot1218a/Business/businessBN/breakingnews-business> [19 December 2002]; John R. Mawdsley, "The Financial Industry Perspective," November 26, 2002, available [Online]: <http://www.iea.org/workshop/amos/mawdsley.pdf> [22 January 2003].

Afterword

1   Letter, Prime Minister Chrétien to John Dielwart, July 24, 2003. [Online]: < http://www.capp.ca/default.asp?V_DOC_ID=766> [28 May 2004].

2   Gary Park, "New Frontiers," *Platt's Oilgram News*, December 1, 2003, p. 3; W.J. Simpson, "A new lease of life; Reserves," *Petroleum Economist*, January 2004, p. 1; Tony Seskus and Joe Parakevas, "Pledge to scrap Kyoto cheered: Oilpatch supports Tory promise," *The Calgary Herald*, June 16, 2004, p. A4.

3   Emma Daly, "Europeans Lagging in Greenhouse Gas Cuts," *The New York Times*, May 7, 2003, p. A8; "EU regulations: Calls to stall Kyoto at odds with political developments," *EIU Newswire*, April 26, 2004.

4   Steven Lee Myers and Andrew C. Revkin, "Russia to Reject Pact on Climate, Putin Adie Says," *The New York Times*, December 3, 2003, p. A1; Sam Fletcher, "Climate's Rocky Road," *Oil and Gas Journal*, March 8, 2004, p. 15; Taylor Hathaway-Zepeda, "Qualifying Kyoto," *Harvard International Review* 26(1) (Spring 2004), p. 30.

5   Alex Scott, "Russia May Rescue Kyoto Protocol," *Chemical Week*, May 16, 2004, p. 17; Guy Chazan, "EU Backs Russia's WTO Entry as Moscow Supports Kyoto Pact," *The Wall Street Journal*, May 24, p. A2;

Editorial, "Quid Pro Kyoto," *The Wall Street Journal*, May 25, 2004, p. A16; "Europe: Mixed Signals; Russian Reform," *The Economist*, May 29, 2004, p. 39.

6   Edward L. Morse and Nawaf Obaid, "The $40-a-Barrel Mistake," *The New York Times*, May 25, 2004. [Online]: <http://www.nytimes.com> [28 May 2004]; Neela Banerjee, "OPEC Raises Quota; Not Much More Oil May Flow," *The New York Times*, June 4, 2004. [Online]: <http://nytimes.com> [4 June 2004]; Erin E. Arvedlund and Jonathan Fuerbringer, "Yukos Warns of Export Halt, Driving Oil Futures to Record," *The New York Times*, July 29, 2004. [Online]: http://nytimes.com; Jad Mouawad, "Oil Prices Set Record Again As Supply Falls," *The New York Times*, August 19, 2004. [Online]: <http://www.nytimes.com> [20 August 2004]; Brad Foss, "World woes push oil to new record," *The Calgary Herald*, August 19, 2004, p. D2.

7   Daniel Yergin, "Imagining a $7-a-Gallon Future," *The New York Times*, April 4, 2004. [Online]: <http://www.nytimes.com> [4 April 2004]; "China becoming world oil market's 'most dynamic factor,'" *Oil & Gas Journal*, June 7, 2004, p. 37; Patricia Van Arnum, "The China Factor: The Wild Card in Global Energy Markets," *Chemical Market Reporter*, June 14, 2004, p. 19; "CERA: U.S. energy supply at 'critical juncture,' IPAA told," *Oil & Gas Journal*, July 5, 2004, p. 34.

8   Steven Strasser (ed.), *The 9/11 Investigations: Staff Reports of the 9/11 Commission* (New York: PublicAffairs Reports, 2004), pp. 68–71, 73.

9   Kate MacNamara, "Oilsands face labour shortage: Project's growth will lead to higher costs, report warns," *The National Post*, October 15, 2003, p. FP6; Canadian Energy Research Institute, "Oil Sands Supply Outlook," March 3, 2004. [Online]: <http://www.newswire.ca/en/webcast/viewEvent.cgi?eventID=747560> [30 July 2004]; Ben Brunnen, "Aboriginal youth hold key to the country's labour shortage," *The Province*, November 4, 2003, p. A18; Gordon Jaremko, "Oilsands growth on pace to double," *The Edmonton Journal*, March 4, 2004, p. G1; Michelle DaCruz, "Oilsands players gird for labour crunch," *The National Post*, March 6, 2004, p. FP5.

INDEX